22°

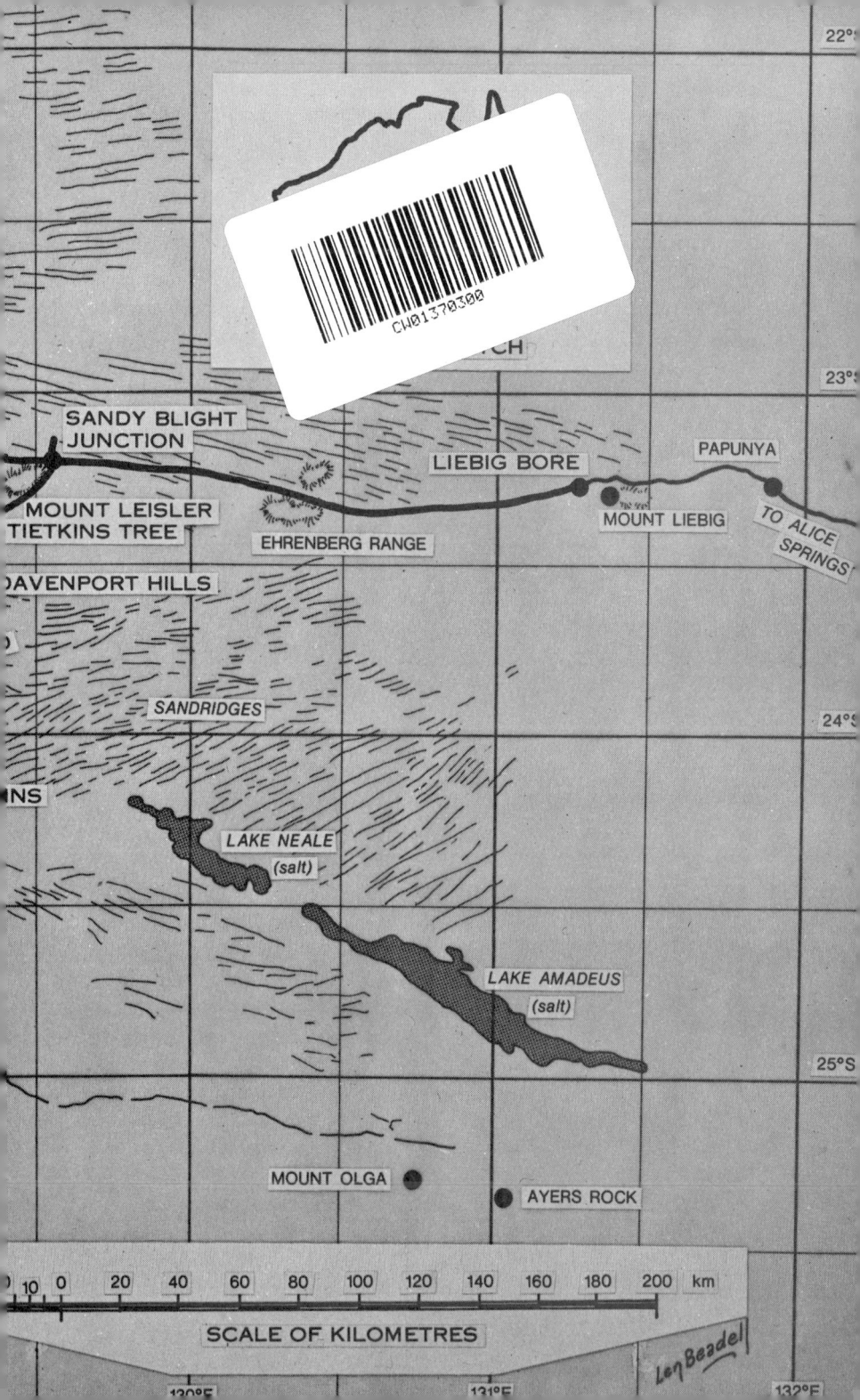

SANDY BLIGHT
JUNCTION

LIEBIG BORE

PAPUNYA

MOUNT LEISLER
TIETKINS TREE

EHRENBERG RANGE

MOUNT LIEBIG

TO ALICE
SPRINGS

DAVENPORT HILLS

23°S

SANDRIDGES

24°S

LAKE NEALE
(salt)

LAKE AMADEUS
(salt)

25°S

MOUNT OLGA

AYERS ROCK

0 10 0   20   40   60   80   100   120   140   160   180   200 km

SCALE OF KILOMETRES

Len Beadell

130°E   131°E   132°E

Len Beadell
Beating About the Bush

LANSDOWNE

Books by Len Beadell:

Too Long in the Bush
Blast the Bush
Bush Bashers
Still in the Bush
Beating About the Bush
Outback Highways (a selection)
End of an Era

Published by Lansdowne Publishing Pty Ltd
Level 1, Argyle Centre,
18 Argyle Street, Sydney NSW 2000, Australia

First published 1976
Reprinted 1989
Reprinted by Lansdowne Publishing Pty Ltd 1994, 1997, 1998

© Copyright: Len Beadell 1976

Wholly designed and typeset in Australia
Printed in Australia by McPherson's Printing Group

National Library of Australia Cataloguing-in-Publication Data

Beadell, Len, 1923–
    Beating about the bush.

    ISBN 1 86302 404 2.

    1. Roads—Australia, Central—Design and construction.
    2. Australia, Central—Description and travel. I. Title.

625.709942

All rights reserved. No part of this publication may be
reproduced, stored in a retrieval system, or transmitted in any
form, or by any means, electronic, mechanical, photocopying,
recording, or otherwise, without the prior written permission
of the publisher.

Contents

1	Hot Water Laid On	7
2	They Don't Make Them Like They Used To	18
3	A Desert Tragedy	29
4	Sandy Blight Junction	40
5	Mount Leisler	50
6	Tietkens' Tree	61
7	The Discovery at Davenport	72
8	An Open-air Workshop	82
9	Bush Bedrooms	93
10	The Border Crossing	105
11	The Discovery of a Desert Tribe	116
12	The Sir Frederick Range	127
13	Roads Do Make It Easier	138
14	Lake Hopkins	150
15	"This 'Arid' Land"	161
16	An Exciting Thought	171
17	Bungabiddy and Home	181

To all those benevolent readers whose eulogistic reviews provoked yet another book

Chapter One

HOT WATER LAID ON

With a startling burst of energy completely unexpected from one of his size, Quinny sprang into the air off the dust heap on which he'd been sitting. His black cloud of flies, having so suddenly lost their target, turned to a less concentrated grey as they dispersed and milled about, and a yell erupted from the still-airborne figure. The loud "Gosh almighty!" broke up the former quiet of the searingly hot spinifex plain where Frank Quinn and I had been relaxing in the scanty shade of a lone mulga tree alongside our recently-made road. Several months earlier I had purposely made the road close to that scraggy tree with a view to its helping some future traveller to boil his billy and have his "dinner camp" in the twelve per cent protection it afforded from the blazing summer sun. As we waited for the rest of our little convoy to catch up with us, we had been idly watching a pall of black smoke boil up out of the shimmering horizon high into the brassy sky, in the direction from which we had just come.

Our new road would eventually extend for 1600 kilometres north-west across Australia, from the Centre to the Indian Ocean. It was to skirt the northern extremity of the Gibson Desert and cut right through the heart of the Great Sandy Desert in Western Australia, finally linking up with the Mid Northern Highway in the vicinity of Marble Bar.

So far we had reached a point about 560 kilometres on our way, but the scorching November sun had given us thoughts of suspending operations until the New Year, leaving the summer inferno of the desert to get along without us. My Gunbarrel Road Construction Party had, as usual, been out in the bush since the previous February, weathering all conditions from raging dust storms to ice in our water tanks. By the end of the year the half-dozen of us were beginning to show signs of tiredness, and so was the battered machinery. The intense heat now magnified any weaknesses normally hidden in the colder months. Radiators boiled and simmered like pressure-cookers throughout the day, and our comparatively meagre water supplies were hard put to keep up with their constant need for "topping-up."

By now Quinny had returned to earth, only to continue his extraordinary behaviour with a bellowed "This is absolutely terrible!" When I asked him what on earth could possibly have brought about this shattering disturbance, he beckoned me to where he was standing in his worn-out, dust-covered sweaty clothes, staring at the rear end of his long-distance supply truck. I left the doubtful shade cast by the tree and brought over my own set of worn-out, dust-covered sweaty clothes, adding my swarm of flies to his. I had stopped my Land Rover in front of his truck, and both vehicles were showing the torture to which we had subjected them since they had last seen a workshop a year ago.

After a minute of giving me the opportunity to discover for myself the cause of his concern by studying me silently from under his battered, oily hat-brim, he asked me incredulously why I couldn't see what was wrong. My examination roved over the back of his truck from the tailboard to the rear tyres, then back to the shiny towbar and spring shackles, and I had to admit I saw nothing amiss. Quinny's hat-brim was pulled down so low over his eyes as to half cover them against the white glare of the sun reflecting off the spinifex tops, but now he cuffed it back on his bald, nut-brown forehead and thundered his distressing question: "Where in the blazes is that great caravan I was pulling?"

At the head of the road to date, just as we were considering calling a temporary cessation to the year's work, the grader had decided to make up our minds for us. Scotty, the operator, was placidly grading the initial cut made by Doug on the bulldozer when a most expensive-sounding crunch from somewhere down in the transmission reverberated through his machine, bringing the whole thing to a grinding halt. After the dust had settled it became quite apparent that the grader would move no more under its own power. Without help, it would have to remain right where it was.

We had already made several contacts with Aborigines in the area, and as they were more than 900 kilometres away from the nearest habitation of any kind they were a wild-looking collection. Five or so at a time, or sometimes one alone, they would come over the sandhills to watch us once they heard the sounds of our activities in their areas. The crippled grader could not be left to their mercy standing out in the desert in the middle of the road throughout the ensuing summer. Our previous experiences had dictated that there wouldn't be much of it left to come back to. In past cases with other engines and machinery, grass, sand, and leaves had been jammed into the diesel tanks, anything bendable had been broken off to use as digging implements in the search for goannas and rabbits, and fuel containers had been bashed open with rocks in the hope that they held drinking-water.

As it was already November, the only course left to us was to tow it behind the bulldozer back along our own newly-made road to where it could be left in safety for repair work. I decided to take it to Giles near the Rawlinson Ranges, a distance of at least 800 kilometres. We had established Giles six years before as a weather station for the Maralinga atomic tests, and it was still in full operation as a remote outpost transmitting weather reports to the Bureau of Meteorology. The initial reason for its inception had ceased completely with the international agreement to call a timely halt to atomic tests in the atmosphere.

Once there, our faithful old machine could be repaired by experts who could fly out the necessary parts to land on the airstrip we had made several years before with that self-same grader. It would then be ready and waiting for us to resume our work in the New Year, with all its levers and fuel lines intact, and minus a diesel tank stuffed full of leaves.

The projected towing operation would simply involve tying the steering-wheel rigid with rope and then hooking the grader behind the slow-lumbering bulldozer. The water trailer normally pulled by the dozer would have to be added to the "train" behind, as that pontoon-type tank would have to be refilled back at Giles for the New Year, and in addition we needed to protect the arc-welding motor, also mounted on its trailer, from the Aborigines and their grass. The bulldozer's average speed was about three kilometres per hour, and at this rate, grinding along at thirty-five kilometres a day, the 800 kilometres could be covered in about a month—if no further problems arose. When a bulldozer is working at bashing down logs and piling up rocks, the wear on component parts like the Caterpillar track-rollers is much less than that caused by the action of "walking" it over long distances. Every two hours for the next four weeks those rollers would have to be greased. But all these things were incidental and almost routine to us. We had been attacking this huge area of untouched and mostly unexplored Central and Western Australia for the past eight years, opening it up with our network of thousands of kilometres of road to provide access for our rocket range and atomic bomb projects at Woomera and Maralinga.

Before starting the return trek we decided to carry on with the next thirty-odd kilometres of road using the bulldozer alone. I had already planned that distance in detail, threading the route through a bad sandhill patch ahead. In the following year, as we continued with the work, I was to be very grateful that we had made this decision.

In the week we spent working on that stretch, our camp remained where it was among a most pleasant stand of desert oak trees. Normally our practice was to move it up daily to the head of the road to conserve petrol. The heavily-laden fuel and service truck, labouring along a soft, newly-made road, managed only a few kilometres to the gallon, and this had a very real effect on our petrol supplies as the service truck needs to attend to the leading bulldozer every day. Those supplies of fuel, water, and rations were brought up to us by Quinny, who was virtually on the move all the time. He constantly had to drive back along our new road to the nearest supply source—a trip which averaged a return distance of 3000 kilometres. Of course, while he was away on these fortnight-long expeditions, we would have moved over a hundred kilometres further on, so Quinny's trips became progressively longer.

Eventually, making most use of the lone bulldozer, Doug and I brought the road through the difficult thirty kilometres of sandhills and, turning the heavy machine around, headed for the camp from which our long haul was to begin. This point was in an area near to some low-lying potential swamp country—if it ever rained in this most isolated region. It so happened that between the time of that memorable turn-around and the time we finally linked the road up to the north-west, a survey party from the National Mapping Council drove along to this temporary terminus. When they reached the end of the bulldozed road they camped, dug a well in the swamp area, and discovered water (of the most awful composition but nevertheless water). They timbered the sides of the well and then covered the top with chopped logs, leaving an opening for a rope bucket. They added to this a most attractive arch of uprights supporting a heavy log crossbar with a squared-off surface on which they carved, in very neat and professional-looking letters, the name "Jupiter Well." In the clear night skies over the desert this large planet stood out prominently from the congregation of other dazzling stars, and so it seemed most appropriate for the new well to be named after it. Jupiter Well has been published on subsequent maps covering the region as such, and it automatically marks the exact location of the beginning of the longest bulldozer "walk" and towing operation in the history of Central Australia.

Back in the camp alongside the sadly immobile grader we lost no time in heading off. Everything had been refuelled, greased, and made ready (not that everything was anything *but* always ready for shifting on a day to day basis), and the bulldozer was slowly reversed, attached to the improvised towbar we had made with the welder, and bolted to the front end of the grader. The cumbersome water trailer had been pushed into position with one of our three trucks and linked up to the towing pin on the grader. As the latter was still standing where it had broken down, pointed in the north-westerly direction of Marble Bar, a huge turning circle had to be negotiated so that we could once again be lined up on the road back to the south-east. It was the most ungainly train imaginable, protesting with groans and shrieks of metal pins and steel tracks, but when all components regained the road in a straight line, Doug was finally able to throttle down and throw out the clutch lever.

We had our mugs of tea and a meal fixed by good old Paul, our cook for many long years, and decided he should slowly follow behind the Caterpillar train, leaving it to set the pace. Rex, in the mechanics truck containing all the main tools and fuel, would come last so that he could lick anyone's wounds on the trip. Quinny in his truck and I in my Land Rover would go on in front and have the meal fires for "dinner camps" and overnight stops going by the time the others lumbered along.

This meant that we had to drive a distance of about ten kilometres and then wait until the dust cloud drew near, giving us the signal to move on a kilometre or so, or to get the fire going, according to the time of day.

The Caterpillar train was not the only one in our little convoy. Paul was towing our refrigerator trailer, which in turn pulled along a Jeep trailer with the blackened stove and iron flue tied over a pair of tables we used for our meals. Iron pegs for an awning to keep off the heat or cold while eating throughout the year, axes, shovels, grill irons, crowbars, and battered water-boilers clanked about as well in Paul's second trailer, but, at the usual speed of six kilometres a day, nothing suffered.

Quinny had to pull behind his truck the old aluminium caravan we had acquired at Woomera several years before, because we wanted Rex to be free to move about as running repairs demanded. I was similarly free of encumbrance in case of emergencies, and the second workshop Land Rover had nothing to drag along either. This vehicle was used by our "cherry picker," who followed the final grading of the road, throwing off the roots and sticks which at times became hooked around behind the grader blade. When the need arose, Eric would also use it to ferry news of breakdowns back to the camp, and Rex would soon appear, complete with his three-tonne fitter's truck, to handle them. A month or so before this, Eric had the misfortune to show signs of a condition in his throat which clearly indicated that expert medical attention was urgently needed, so he had temporarily left our party to return to civilisation on one of Quinny's ration trips. On this epic trip, however, his workshop Rover was being driven by Scotty, who was at the present moment for once without his precious grader.

We were only four days away from the terminus camp when Quinny made his outburst, and as soon as he asked his very obvious question concerning the whereabouts of his big trailer, I could start sharing in his anxiety. Up till then it had been remarkable how we just sat there for so long without noticing such a really conspicuous calamity, but now we set about planning how to recover the lost vehicle. Of course, the only thing to do was to retrace our tracks back along the road until we discovered it. Even if Rex had already found it, a repair job would probably be needed on the draw-bar and he would need help. Meanwhile the slow bulldozer train could trundle on regardless.

As we talked it over we glanced back across the wide expanse of hot spinifex in the direction from which we had come, but no trace of the caravan could be seen in the mirage. The black cloud of smoke was still in evidence about six kilometres away and we both decided it was coming from one of the huge piles of dry spinifex and dirt left by the bulldozer on our way out. When making the initial cuts with the heavy blade the rubbish soon heaps up to higher than the radiator-top on the machine and has to be "dumped." Pulling slowly off the course of the road, Doug would push it clear and lift the blade over it, driving the great steel tracks over the top to flatten down the loose clumps so that they wouldn't blow back over the road. These mounds of dumped rubbish are to be seen at short intervals beside any new road cut through scrub or unbroken ground. In our case, to help keep the loose bits of grass and debris from moving about with the wind, one of the team would often throw in a match to burn the lot down. There were obviously too many to worry about over hundreds of kilometres of new road, but at a brief re-greasing stop a near-by pile would sometimes idly be lit.

As we turned our attention from the billowing volumes of black smoke to the bare trailer-less space behind Quinny's truck, another sight caught our eyes. A large plume of fine dust was forming between us and the smoke, starting as a pinpoint and swelling to an enormous red cloud. A vehicle was coming towards us at the fastest speed the soft new road would allow. The little workshop Land Rover was the only thing in our snail-like convoy capable of such a pace, and we both thought Scotty was coming to advise us of the fact that a stray caravan was lying dormant in the desert.

Postponing our turn-around until his arrival, we were all set to listen to his news with feigned great surprise before retracing our tracks for its recovery. Eventually the little Rover ploughed to a stop in a swirl of bulldust which temporarily obliterated it, and we heard Scotty yell out from somewhere in the miniature dust storm: "*The ration truck's gorn!*"

By this time we could actually see him through the subsiding haze, and no further elaboration was needed to boost the full impact of what he had said. The real cause of the black smoke became suddenly all too apparent. How could a harmless truck full of our food supplies catch on fire, even though the soaring temperature of the day made everything too hot to touch? As we were still over 600 kilometres from anywhere, I thought that this was definitely going to break the monotony.

Without further unnecessary conversation, I leapt for my Land Rover and we headed back along the road as fast as we could, leaving Quinny to turn slowly around and follow us to at least as far as the caravan, wherever that was. As I pushed the vehicle along the smooth powdery road the thought occurred to me that, since the previous ten minutes, things could only improve. After travelling about three kilometres I came to the caravan, still upright, but with its broken towbar ploughed into the ground at an angle to its normal direction of travel. A few minutes ago the loss of the caravan had been the great catastrophe, but it now receded into the background as being quite unimportant. We drove around it, hardly giving it a second glance, and soon the ration truck became visible at the source of the huge mushrooming pillar of black smoke.

We couldn't approach too closely at that moment because exploding tins of stew and bottles of tomato sauce were showering the area spasmodically with carrots and shattered glass, so I stood well clear beside Paul, just staring at it all. Without pointing out to me that his truck was slowly melting in front of his eyes, a fact he deduced I could see for myself, he did casually remark in a quiet, awestruck whisper that this was terrible. Agreeing with him readily, I volunteered the opinion, as another carton of tins blew up, that you could say that with a reasonable amount of truth.

The big tyres were spouting globules of molten rubber, leaving blazing trails behind each pellet, and beyond the conflagration around the red-hot rims, the once upswept thick assembly of heavy springs was to be seen draped downward at the shackles like a foot-thick bar of toffee. The cabin where

Paul had been sitting less than hour before, quietly driving along the road, was a white-hot shell with flames still fighting to get out of the open windows, and the windscreen had softened so that it lay draped out over the bonnet like a tablecloth. The little fridge trailer was also alight at every burnable point: engine belts, canopy, tyres, and rubber seals. The ribs which once held the canvas covering over the food supplies made the whole thing look like some large prehistoric skeleton which had been barbecued by eager hungry cavemen.

When the worst of the minor explosions had eased off, I took my camera, shielded my face with a hat, and ventured closer, hoping to try and record some of this to show the yet unsuspecting transport officers at headquarters. Even as I focused the instrument a tin of pea soup burst its seams, splattering me all over with thick pale green liquid, but I carried on with the pictures after wiping a blob of soup off the lens with a piece of what was left of the pocket-lining in my old shorts.

The rest of the party had by now arrived and Doug was brought back from his bulldozer train to share in the spectacle. Rex saw the engine he had faithfully serviced for so many months lying in among a blistered array of panelling as just a shapeless block of useless iron. We did see one of the wheels as yet not actually on fire, and although it was too hot to handle we attempted to take it off with a wheel spanner to save it and use it to replace a damaged one on our other truck. Jacking it up, with one hand protecting his face, Rex managed to loosen the big nuts and free the rim from the simmering brakedrum, attaching a cable to it when it fell clear. A Land Rover dragged it away across the spinifex to cool, and that became the only item salvaged. I did hear a bubbling sound coming from the 1400-litre water-tank installed behind the cabin and it occurred to me we could at least make a cup of tea. The tap had melted so I shot a hole through the side of the tank with my revolver and, finding some tea-leaves among the wreck, filled a billy from the boiling geyser. The sugar in the once-new bulk container had melted, but some could be obtained with a shovel after digging a hole into it with a geological

TOP: Our camp near Jupiter Well, four days before the ration truck disaster. It was from here that our marathon towing operation started. BOTTOM: "They don't make them like they used to." The burnt-out remains of the truck

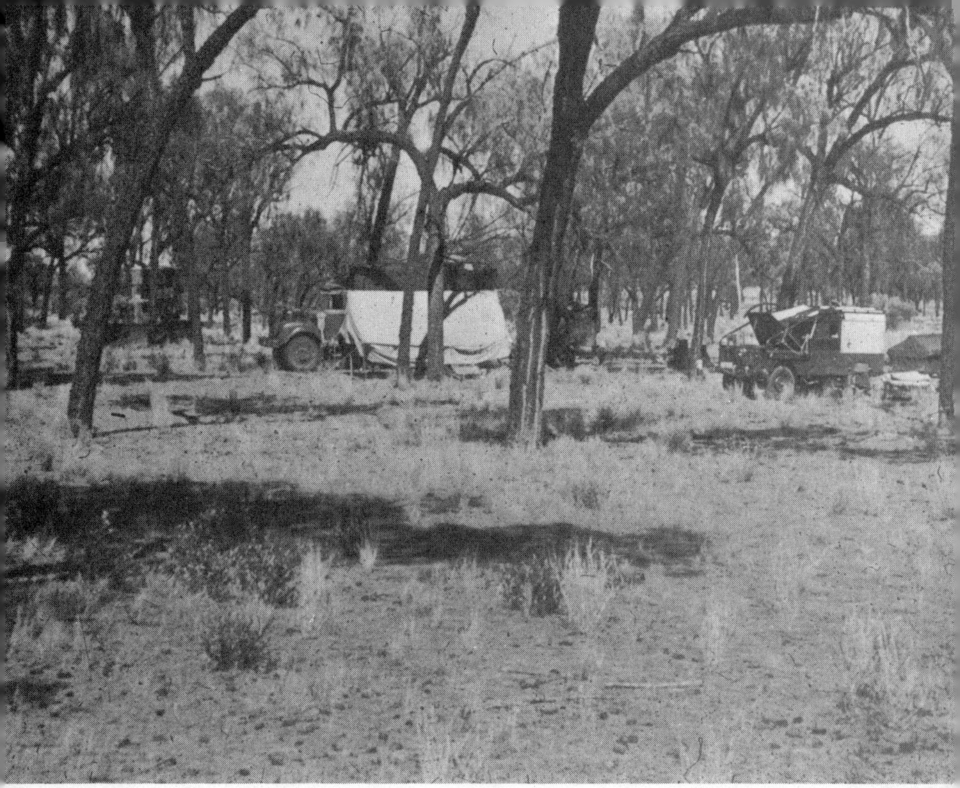

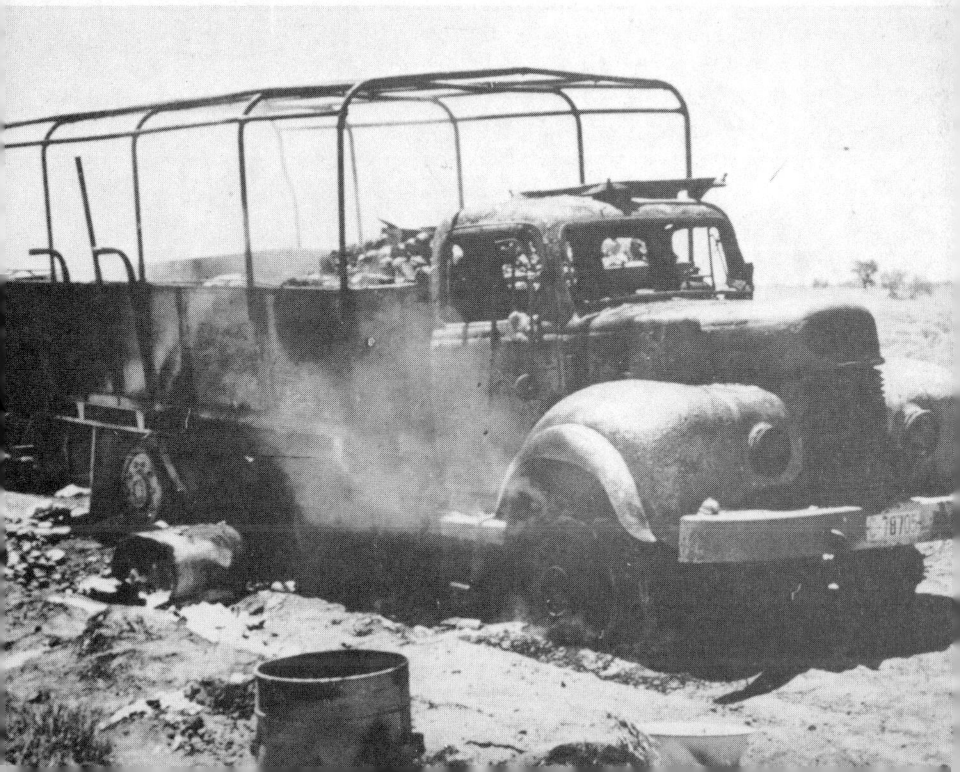

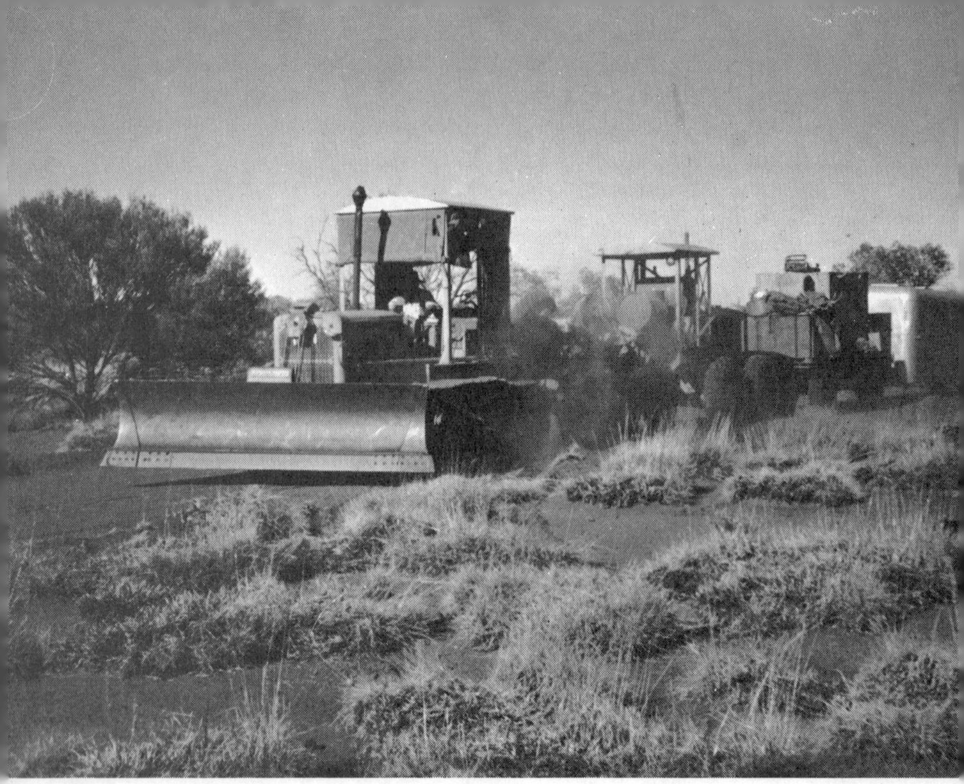

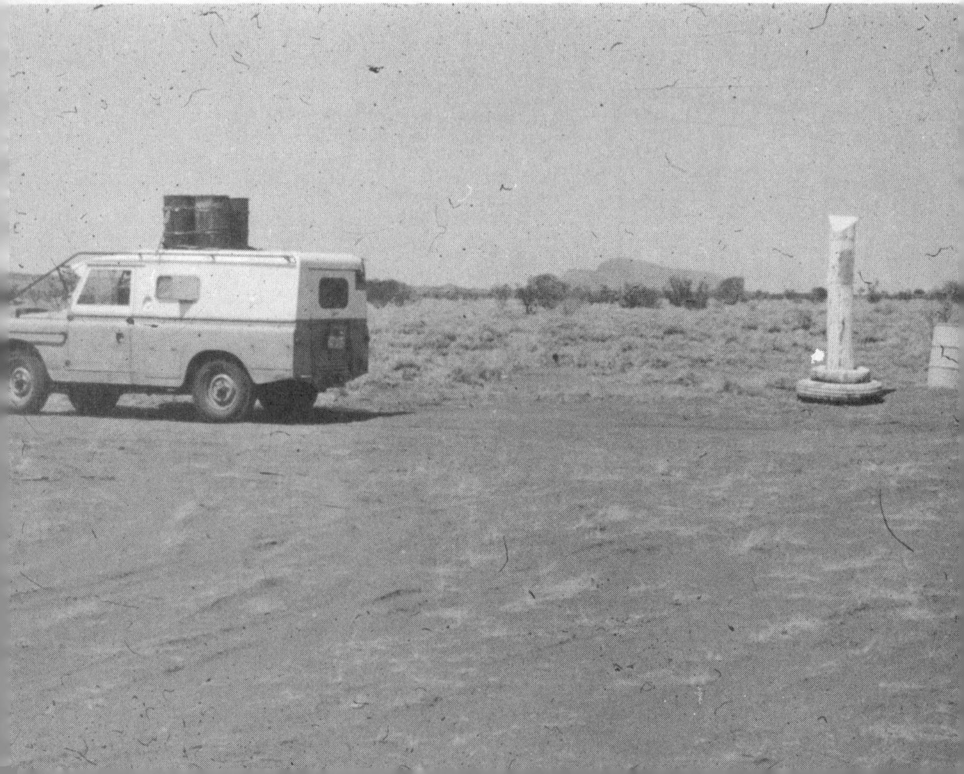

hammer. As we poured out the mugs of tea, each of us was asked politely if he preferred one shovelful or two as a sweetener.

I asked Paul to tell us what had happened just prior to the discovery of the fire, and he explained that as he drove slowly along he became aware of a lot of popping and crashing sounds from his load. When he looked out of his window around to the body of the truck he saw that everything behind the cabin was a holocaust of leaping flames and smoke, so he thought to himself that perhaps he should amble away from it before it exploded. Any extinguisher we carried was by then seen to be useless against the hold the fire had on the truck.

Later, as he stood beyond the reach of the flying debris, Paul was heard to remark sleepily that there was something you don't often see.

TOP: The bulldozer train trundling slowly back down our newly-made road. BOTTOM: Sandy Blight Junction, of painful memory. Mount Strickland is just visible in the background

Chapter Two

THEY DON'T MAKE THEM LIKE THEY USED TO

When the flames and smoke eventually abated and the explosions of the food-tins became less frequent, the burnt-out shell of our ration truck made the desert look quite untidy, squatting as it was squarely in the middle of our brand-new road. After Paul decided to bale out, it had ground to a stop within a centimetre of where he had depressed the brake pedal, and now the huge glowering hulk was blocking the path of any subsequent traveller. As yet no one had even heard of this new road of ours, much less driven on it, and the long straight ribbon of this section which I'd taken so much trouble to plan stretched out in either direction to the horizon. Anyone intrepid enough to make a future pilgrimage to these parts would have to drive around the obstacle, a simple enough undertaking in itself, with unlimited room for the manoeuvre; but unless we moved it I could visualise a deep set of wheel ruts breaking up the neat continuity of the otherwise impressive straight line.

With this thought in mind, we had Quinny position his truck off to one side with its big winch facing the scene of destruction. A long steel cable was attached to the still-hot bumper bar, and as the strain was applied the wreck began to move. Because the front wheel had been salvaged, the dead weight scooped up a mound of dirt in the first metre, which distance was all it travelled before the winch cogs smashed. The equipment which had served us unfailingly throughout the year was slowly falling to pieces in front of our eyes. After we had unhooked the refrigerator trailer and added another truck to the operation, the simmering heap was finally pulled clear and parked forever neatly alongside the road, where it remains to this day.

Rex had been studying the extent of the damage to the fridge trailer and gave his opinion that it could possibly be repaired in the field. Although everything burnable had gone, the engine and compressor were still recognisable. The towbar which had been joined on to the main source of the fire was still too hot to handle, but eventually it was hooked on to Rex's truck for further study at our night's camp. The only piece of personal belonging saved was Paul's swag roll which he had managed to untie from its spot on the front of his truck and drag clear before the petrol tanks exploded.

No food could be saved. Only a few hours earlier the sad-looking pile of hot iron had housed all our supplies, and this left the six of us out in the desert without a thing to eat. Actually one or two stray tins of meat were to be found in other vehicles, and I always carried a few items to make my Rover independent from the others during my never-ending succession of lone reconnaissance survey expeditions, so these we pooled. Together they added up to not much more than a couple of meals for the entire party, so a plan of action had to be formulated on the spot. None of us was in the least concerned, as our whole lives for the past decade had consisted of facing up to and dealing with problems and large-scale difficulties. It was, as Quinny so gleefully put it, the best entertainment we'd had for a year. The grader was broken up internally, the winch smashed, the van had ploughed into the red dirt, and we were now minus a complete truck, but otherwise we were perfectly all right.

The cause of the fire was impossible to ascertain. Perhaps some electrical wire had frayed through and, helped by a cargo already seething from the furnace-like sun beating down on our little cavalcade, a spark had done the rest. A scorching head wind would have fanned and cultivated the blaze, blowing the resulting smoke away from the cabin until, by the time Paul discovered it, it was past extinguishing.

Our attention was now directed towards the stranded caravan. A repair job with the mobile welder was needed on its towbar before it could be moved, so Rex, complete with the hottest refrigerator trailer in Australia, trundled off towards it. Using the welder and pieces of broken spring leaves, the towing apparatus was soon repaired and coupled to the bulldozer train. That most wonderful Cat machine now effortlessly hauled behind it the twelve-tonne grader, the huge steel pontoon water-tank on its four-wheel trailer, and the newly-recovered caravan, while Doug at the controls passed the time by reading. Once in motion, the steel tracks kept their own straight line and remained in the direction they were aimed at. Such a short distance was covered in so long a time that there wasn't anything else for Doug to do, apart from listening to the shrieks and groans of protesting steel as it scraped and banged together. Every two hours the whole convoy had to be stopped while Rex pressure-greased the tortured track rollers upon which everything relied.

The very tree under which Quinny and I had crouched earlier, idly watching the column of smoke in the distance, turned out to be our overnight campsite instead of just a proposed lunch stop.

With the dozer train still a long way off, the rest of us reached that spot to begin the appraisal of our situation and look harder at the possibility of getting the fridge to function once more. My next duty was to call up our base transmitting station at Woomera on my mobile radio transceiver and relate the irregular happenings of that day. I only used this system when needed, instead of keeping a regular schedule, so Ron, the operator, received his customary shock when I gave my call sign VL 5BW8.

The reply coming clearly over my loudspeaker indicated that although he could make out the call sign, Ron could hear little else through the "hash" caused by the heat and the great distance between our little battery-operated radio and his powerful complex. As against his huge antenna system, ours made do with a whip aerial. Mostly this worked extraordinarily well on voice, and when the radio telephone switch didn't produce results, I could always change to carrier wave and use the morse key as Ron was an expert reader. I could transmit morse reasonably respectably, but receiving was another matter. However, that didn't affect us at all as I could always hear his voice signals.

So the story slowly unfolded, with Ron on R.T. and me reporting the expensive mishap on C.W. I asked him to relay the news a further 500 kilometres by teletype to our Weapons Research Establishment headquarters near Adelaide. Our ration truck had unfortunately melted, but we were coping adequately with the situation. After I included the current temperature of 45°C the reply eventually came back that in the opinion of the transport section officers, we got it rough out there in the desert. My reply, transmitted laboriously on the key, wafted across the blistering Central Australian expanses to state that they just don't make them like they used to.

The plan which had been developing in my mind needed to be put into effect without further ado. Quinny and Scotty, the graderless driver, were to travel non-stop in Quinny's truck to

Giles, stock up with food and water, and return to the convoy which in the meantime would still be able to creep along. Taking turns to drive, with no stops other than refuelling, the whole return distance of 1300 kilometres could be made before the members of the trek fainted from hunger. Before signalling "over and out" to Ron, I asked him to contact the little weather station to prepare them for the impending raid.

I would also have to make a trip of 1500 kilometres to Alice Springs and back for various items lost in the fire — equipment for Paul's personal use, a supply of fresh meat, and certain tools necessary to complete the remaining 600-odd kilometres of our towing operation. This had been somewhat interrupted, but the actual onward movement of the slowest part of the convoy, namely the bulldozer, had so far been barely affected. If our two long trips went without incident, the overall time taken by the mammoth tow might then be lengthened only by the half day's excitement. Under the circumstances this amount of time would be negligible.

Within minutes of passing on the plan, Quinny and Scotty were off and away in the former's three-tonne vehicle, and in a few seconds they were lost in a billowing cloud of red dust whipped up from the surface of our new road. Very little discussion or arranging was needed with such men. They were as thoroughly able and reliable in all emergencies as they were in routine hard work, and we knew they could handle unforeseen mishaps as efficiently and well as any bushmen in Australia. Frank Quinn once drove a team of camels in the outback long before motors came his way, and Scotty, who came from the country, was as practical and experienced as anyone of my acquaintance.

Not wishing to delay either, I decided to head off to the Alice. My absence would leave a total of three less mouths for the meagre food "pool" to feed. Our spare transmitter box would be left as a contact between the camp and me, and Rex could inform me of any further requirement resulting from our losses while I was away. At a given time each day we would call each other, and if Rex was subsequently successful in mending the fridge, among other things, he could let me know. My back-loading would depend on his information. If the little box was made to freeze again, I could return with a large supply of deep-frozen fresh meat wrapped in bags of dry ice and a camp-sheet, hoping it would survive the long hot trip back until Paul could install it in the cold once more. Of course, if the engine and compressor had been burnt beyond repair, that plan would be futile, and we'd carry on with the tins from Giles. All this would in no way impede Doug's progress with his train.

A quarter of an hour after the first departure, I added my smaller cloud of dust to the desert's night air. We would both be travelling on the one and only road back for 170 kilometres to a point I had named "Sandy Blight Junction," where our paths would split up. The truck would turn south to Giles, and I would continue on to the east and Alice Springs.

With their head start Quinny and Scotty had made the junction before I caught up with them. When I arrived I could see their fresh tracks turning south around the big signpost log we had installed there with our aluminium plate nailed on to the squared-off section. The details of the distances for the other directions were just as I had hammered them on with my set of alphabet punches, but of course the one pointing to the west would have to wait, as we were still in the process of making the road. Without stopping I carried on straight ahead to make as many kilometres as I could before tiredness claimed me after the day's out-of-the-ordinary events.

The road running easterly from Sandy Blight Junction joined on to the existing track to Alice Springs at a windmill and stock tank named Liebig Bore, a distance of 180 kilometres. I had made this link mainly so that we would have access to the Alice for the supplies we would need constantly as we worked our way north-west towards the "Eighty Mile Beach" on the Indian Ocean. Parts could be readily obtained for the bulldozer and grader as required. It was also our nearest railhead, even though the return "shopping" trip would ultimately reach 2500 kilometres as we progressed further into the Great Sandy Desert.

Liebig Bore was the furthest western watering-point for cattle belonging to the Aboriginal settlement of Papunya, eighty kilometres away from it. Papunya was in turn linked to the Alice by a graded road. The connection to my road system was an obvious necessity, although much in the way of rations would come by Bristol Freighter to Giles, from where they were periodically collected by Quinny. Mount Liebig towered close to the bore, forming the western extremity of the MacDonnell and Amunurunga ranges in the Haast Bluff region. The whole section of that country was among the most picturesque in Central Australia.

A well-used station track joined Papunya to Liebig Bore, and it was here the existing road system terminated. Our own system therefore just had to be linked up to give access from Alice Springs clear through to the Marble Bar, Broome, and Port Hedland areas. Midway between Sandy Blight Junction and Liebig Bore my road passed close by the Ehrenberg Range, where Harold Lasseter had made a supply dump for himself a third of a century before. It was from here that he made his now famous lunge south to the Petermann Ranges and his elusive reef of gold; a discovery so controversial that I would not dare add my own beliefs to the volumes written about it.

At dead of night I finally stopped the Land Rover. I was now within a kilometre of where Lasseter had left his canvas-covered cache and the plywood noticeboard requesting anyone finding it not to interfere, as his life depended on it. The hot, absolutely quiet bush surrounded my lonely camp as I lay down in my swag, as tired as any mortal could become, and fell asleep within seconds. In the early hours of the morning I became aware of a disturbance of my blanket, accompanied by a noise which was enough to make me fully awake immediately. A violent wind had sprung up, whipping dust and mulga trees into a frenzy. It was only an hour to daylight, so I managed to roll my swag and escape into the cabin of the Land Rover. The force of the gale rocked the little vehicle back and forth on its springs, and when I turned on the headlights the beams struggled out into the wall of dirt like two yellow posts and were lost before illuminating anything. As I obviously wasn't able to drive on for the time being I just crouched there and resumed a "closed eyes" position again, draped over the steering wheel.

Surprisingly, I did sleep for a time, because the next view I had through the windscreen indicated that daybreak had arrived and some little section of my road ahead was actually visible. The fury of the dust storm had not abated, but at least I could crawl the Rover along at a snail's pace for something to do. As the sun rose, the distance to be seen ahead extended with every minute and eventually the orange orb of light showed itself through the haze. It was so dim that I was capable of looking at it easily without discomfort to my red-rimmed eyes, and it remained so until mid-morning; but as long as I could travel, neither the sun nor the storm affected me in the least.

When carrying out the initial survey for this road I had aimed one long straight section right towards the spectacular peak of Mount Liebig itself. I hoped future travellers from Sandy Blight Junction would marvel at the grandeur of it, framed squarely in the bulldozed clearing through the thick scrub forming the road edges. Now, as one of the first to drive back along my own road, I could see only dust. The knowledge that it was there had to be good enough for me when I reached the beginning of that long straight stretch.

Eventually I arrived at the signpost I had erected near the bore. The aluminium plate on the post was intended to reassure anybody heading off to the west along the new road that they were on their way correctly, and to give them what distances lay in store for them ahead, as other help of any kind ceased at the Papunya settlement. In this case it reassured me that I had almost reached the bore, with only eighty kilometres left before coming to the superintendent's house where I could gain a temporary respite out of the wind. I hadn't eaten anything since the morning before the fire, and perhaps Mrs Snook there might have a billy on her stove. The intense heat made the thought of eating not so attractive, but I was feeling peckish to say the least as I drove slowly past the bore.

I had made the road well clear of the actual windmill, tank, and trough so that traffic would not disturb any stock about, and as I waded across the dust-filled, grimy opening surrounding the bore, I faintly made out the shape of the enormous black water-tank and the windmill tower above it. The fan blades of the mill were invisible, partly because of the dust, but mainly because the wind was spinning them at supersonic speed.

Continuing on towards Papunya, I came upon the blurred outline of a station truck drooping sadly down on one corner with a wheel and attached section of a broken axle lying alongside in the swirling dust. I drove around it, having seen it in front of me on the track just in time through the storm. The roads hereabouts were gradually becoming blocked by crippled vehicles, I thought.

At last, by mid-morning, the Aborigines' camps and outbuildings at the Papunya settlement began to show themselves, and after battling with the iron gate I finally pulled up outside superintendent Snook's house which was perched on top of an array of high poles.

Mr Snook came down to greet me, and at last, for the first time in nearly half a year, I was installed in a house with walls to keep the raging wind out. My partly-unlaced hobnailed boots looked extremely out of place on the sitting-room carpet and my ragged dusty shorts made the lounge chair protest as I sat down. As I had visited the house before I asked where the other members of the family were.

Apparently Mr Snook's wife had been periodically plagued by an agonising toothache but had steadfastly declined to make the 240-kilometre trip to Alice Springs for treatment. Every time, just when the family had succeeded in persuading her to go, the ache had mysteriously but temporarily disappeared. This wild morning of my arrival was one of the times the ache was upon her in full force, and she was in another room, oblivious of the cyclone outside, and tended by her daughter Gillian. The last time I had called into Papunya, the worst of the pain had passed, and although she knew that in my Land Rover there was an ominous padded dust-proof box containing forceps, hypodermic syringe, and anaesthetic, she had politely rejected my offer of an on-the-spot extraction. She added that if I had happened along the day before she couldn't have resisted the prospect of immediate relief, but her "Dutch courage" had waned with the diminishing pain.

This time was another matter. I began to sense the prospect of an operation even before Mrs Snook, having gathered that the current visitor was myself complete with dust-proof box, had stumbled out of hiding. Absolutely no inveigling was needed as she was now guided to the seat which was soon to become a dentist's chair. I pushed my way back down the steps through the dust and wind and returned in a hurry with my dreaded box in its canvas cover, forcing the door shut again to keep the stinging dust out of the lounge-turned-waiting room.

Out of sight of the patient, as recommended by the manual installed in the lid, I spread out the white cloth and instruments while Gillian put a pot of rainwater on the stove to boil. As the offending tooth was an upper left molar, I selected an appropriate set of forceps from the gruesome bundle and a needle suitable for injecting the anaesthetic with the cartridge-type syringe. Cottonwool tweezers and all the rest of the gleaming paraphernalia were soon bubbling in the clean boiling water while I reassured Mrs Snook that painless extractions had already accounted for over twenty notches on my forceps, and she had everything to gain by consenting to the impending operation. I had had a good teacher (Dr Bruce Dunstan, formerly of Woomera) for my "Desert Dentistry" course, and the best equipment had been bought in by the quivering clerk in our headquarters purchasing section.

It wasn't long before I was the one who needed reassuring, after I had heard my patient's faltering description of her last extraction in England. The roots of her tooth were found to be almost tied in a reef knot, and a major bone operation was needed for complete removal. I thanked her for the assistance this information would give me, and after the prescribed time for sterilisation had elapsed I raised the instruments out of the boiling water by means of the cottonwool tweezers, the handles of which I had left out in the cooler air.

Just at this stage a yell came from somewhere out in the storm that someone was calling the Snooks on the flying doctor transceiver installed in a radio shed near the house, asking about some roofing nails which had gone astray in the mail. A reply shouted through the wind from the "surgery" indicated that the person must stand by for a while. In a quieter voice addressed to those present in the room, Gillian stated positively that she wouldn't miss this for the world.

Only the day before, I had been hundreds of kilometres out west on the spinifex plains being entertained by the spectacle of our blazing ration truck; and now, as I installed the needle in the syringe and loaded the anaesthetic cartridge, I wondered when this latest series of events was going to end. Everybody stayed at a respectful distance as I approached the reclining victim with the long, shiny needle poised.

Chapter Three

A DESERT TRAGEDY

The tooth was a maxillary or upper first molar, so the first thing to be done was to demonstrate the mucobuccal fold which indicated the point of puncture for the required supraperiosteal injection. This was done by drawing downward on the mobile mucous membrane in the maxilla, and the resulting line it made with the firm mucoperiosteum would be my centre of concentration. The book had gone on to say that this should be supplemented by a posterior palatine injection placed halfway from the gingival border to the mid-line, mesial to the third molar. Together they would anaesthetise the distal and palatal roots of the tooth. So, aiming the syringe vertically and expelling the air with a drop of liquid, I began to work, forgetting for the time being the howling wind outside and the original purpose of my trip.

The anaesthetic in the cartridge was zylocaine mixed with Epinephrine, which latter prolonged the numbing effect, and only a percentage of the 1.8 millilitres of the contents was to be used in this injection. The remainder was to be deposited at the posterior superior alveolar nerve in the palatine region. Drop by drop in advance of the needle I injected distally and upward past the mucobuccal fold to the apex of the roots. The manual certainly set it out clearly, and as I began the palatine injection I noticed that no ballooning of the tissues resulted.

Mrs Snook didn't even know what was happening, because at that stage the ache was still in evidence, but in several minutes that marvellous anaesthetic worked to give her the first relief she'd had for days. Then came the time to test the tissues around the tooth until eventually no feeling was left and the extraction could begin.

The upper molar forceps are perfectly shaped for both gripping the tooth and being gripped by the operator, and as I took them up in such a way as to make them invisible to the patient, I remembered all that Bruce had explained in his Woomera clinic. The beaks of the forceps had to be pressed past the gingival border or gum-line to concentrate their action on the roots, bypassing the crown of the tooth completely. I shut out from my mind Mrs Snook's description of her last visit to the dentist as I began the gentle but firm loosening action. Although an extreme amount of effort is needed to break the roots free from the bone, this must be concentrated to very little movement, and soon the muscles on my forearm and wrist were being used to the full. Gradually I began to detect a slight movement of the tooth independent from the maxillary bone, and soon it was free enough to be released altogether. I had always concluded that after that first movement it was almost all over, and sure enough, with the fingers of my other hand placed either side of the gum, the source of many months' torture was soon out in mid-air, still held by the now stained forceps.

Throughout the previous few moments, as is the case with any such patient, Mrs Snook's eyes had been screwed shut, but now that I was clear she ventured to ask tentatively how things were going. I couldn't believe my ears; she actually didn't know the tooth was out. After I had informed her that it was all finished she asked to see the object which had caused her so much misery. I dropped it with a metallic clang into the kidney dish, and then proffered a glass accompanied by the most professional-sounding "Rinse, please" I could muster. In no time my patient was proudly acquainting all and sundry with the information that she had just had a tooth out and there was really nothing to it.

I had delayed long enough, and so after repacking the dust-proof box and installing it back in the Rover I plunged once more into the swirling dust. Somehow the subject of lunch hadn't cropped up, what with one thing and another; and although I still hadn't eaten a thing since leaving the fiery truck, my thoughts didn't seem to include food. Immediately following the operation we had all had a mug of tea, and during this time I happened to hear over their radio that a complete fire ban had been declared throughout the State, with a heavy penalty for offenders. I couldn't help visualising a poor unfortunate officer, complete with waterbag, struggling out west of Sandy Blight Junction through the dust storm to collect the fine at the ration truck.

As I camped in the scrub fifty kilometres from Alice Springs, I began to sense an easing in the force of the wind, and early in the night it ceased altogether. Twinkling stars appeared, although there was no moon, and it felt wonderful to lie on my swag roll in the quiet bush, with the incessant noise of the wind completely replaced by that of the empty rumblings inside me. I could still feel the twinges in my right forearm but at least it was for a good cause.

On arrival in Alice Springs the following morning, I drove straight to the centre of the dry, sandy expanse known as the Todd River and set up my aerial for a contact with my camp. The prearranged time was approaching and I was soon sending out my call sign, hoping the spare transceiver Rex would be using still worked. Although it was supposed to happen, I was still surprised to receive an immediate answer from the boys, telling me the call sign was plain to hear. It followed that they had been successful in actually repairing the fridge's engine and compressor to a state where it again caused the box to freeze, and everything was as I had left it, with the bulldozer train forty kilometres further along the way. Scotty and Frank Quinn had not yet returned with their load of rations and water, but they were still all right, even if they were becoming a little peckish. Actually I suspect they had eaten more than I since I headed off two days before, so, assuring them I would be on my way back in a day with the frozen meat supply, I packed away the aerial and went in search of something to eat.

After organising the butcher across the dry creek to prepare the meat for a long hot trip, and buying in some personal items needed by Paul and the others, I called into the police station. My gun licence would be more in order if it were endorsed to allow me to have the weapon in my possession in the Northern Territory, a fact which had already been mentioned to me by the police sergeant in Woomera. I discovered as I dropped my revolver on his desk that the Alice Springs inspector, McKinnon, already knew of me and our activities out west. However, he ignored the formalities for the time being and turned his attention to the revolver. It was too good an opportunity to miss, he said, as he carried it outside for some target practice and aimed across the street to where some empty cans lay in a heap. I really thought he was going to blaze away, turning the town into one straight out of the Wild West, but eventually he lowered the gun, muttering that in the old days he would have tested it by firing out of the window without moving from his desk.

While I camped overnight in the creek-bed the meat had time to turn solid in the butcher's deep freeze, and first

Straight as a gunbarrel, our road runs towards the magnificent peak of Mount Liebig. The landmark is shown here at a distance of about eight kilometres from where the photograph was taken

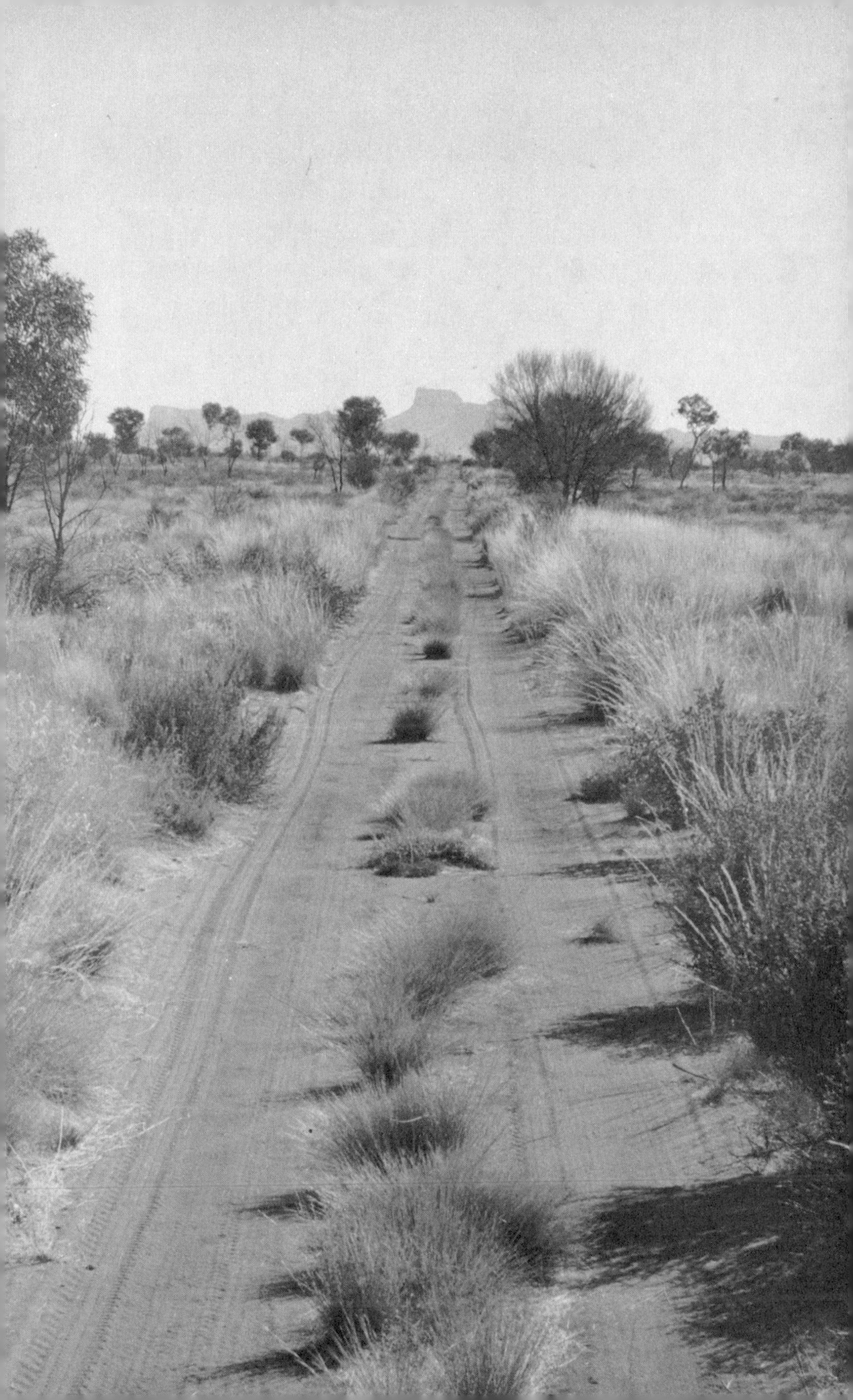

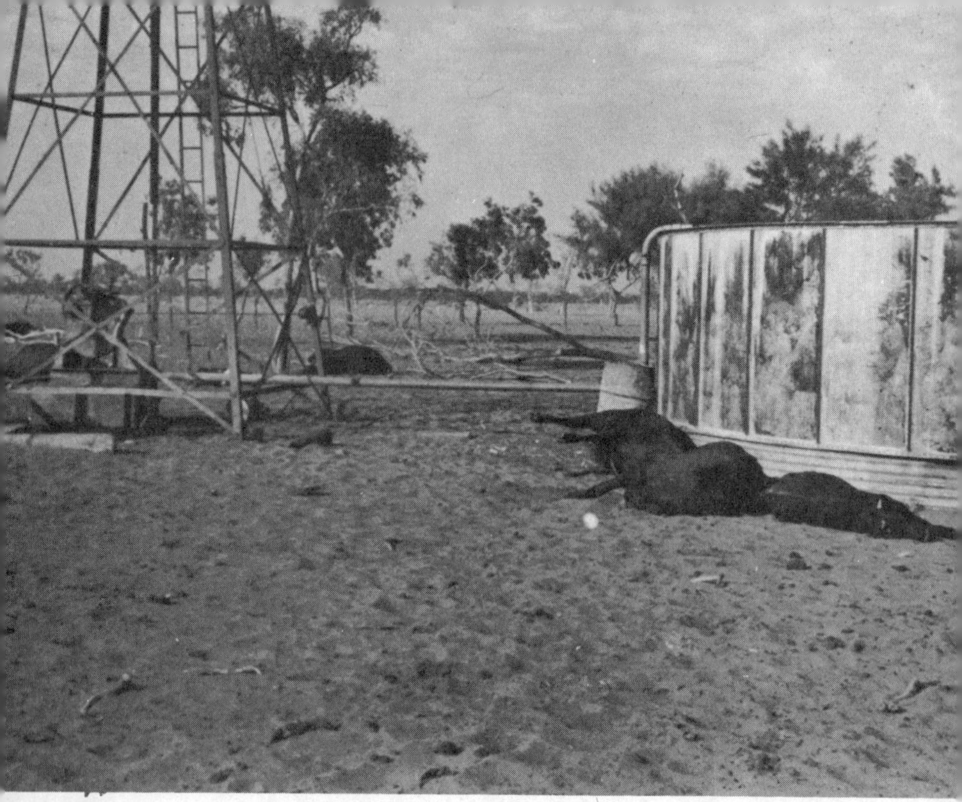

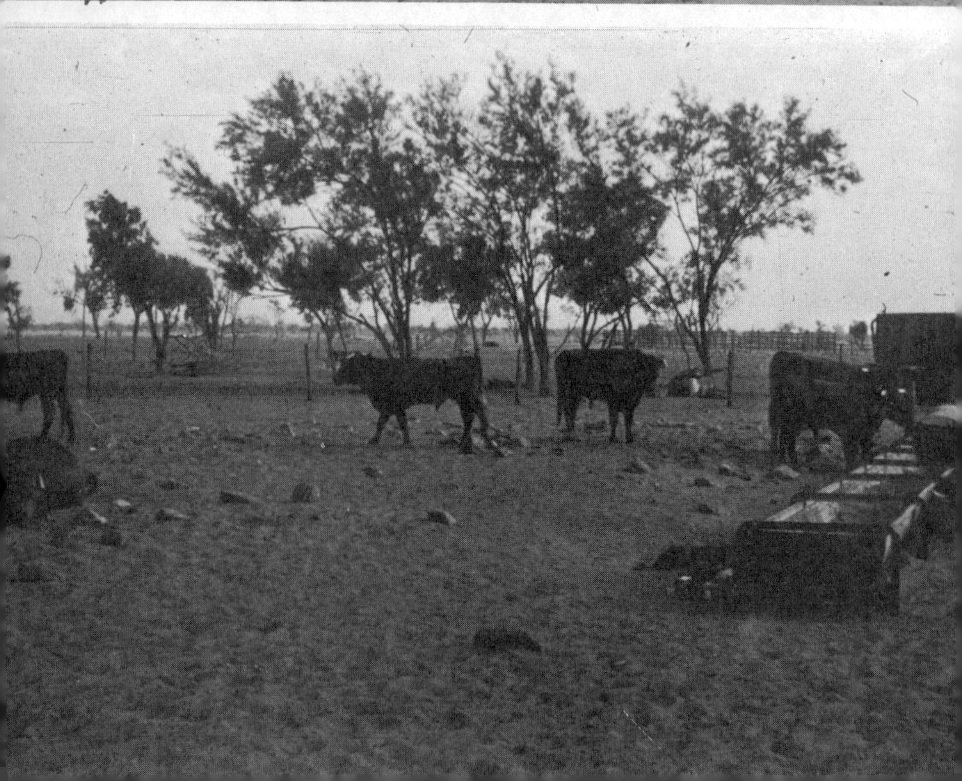

thing in the morning saw us rolling it up in a canvas campsheet together with loads of dry ice. With this huge bundle roped to the roof of my Rover, I pointed the radiator to the west and headed off. I had about 600 kilometres to go and had hopes of arriving back at the camp in the early hours of the following morning when the meat would still be frozen and the transfer to our refrigerator box could be achieved before the flies knew about it. A further tragic interruption was to throw that plan to the winds.

During the trip to Papunya the Central Australian November sun beat down mercilessly, and at any waterbag stops I made, the armies of flies had taken over in force from the stinging dust of a day ago. After finally pulling up outside the Snooks' house, I found that quite a different atmosphere now surrounded the place. The family hurried down to greet me and this time the first thought was lunch. My ex-patient was in exuberant spirits. However, not long after we had all sat down to eat, there was a knock on their door and when it was opened the forlorn figure of the station cook was to be seen clutching a small cloth bag. His usual smile had gone and a toothless mouth informed us that he had just broken his upper dentures. The word had obviously spread. As my repertoire, supplemented by the tuition of the Alice Springs expert, Ray Meldrum, included the repair of such useful items we asked him to leave his little white bundle and we would attend to it.

After our meal the dust-proof box was again opened, and the dental plaster, sticky wax, and jars of Paladur powder and liquid were laid out together with a Dappins mixing glass. In a reasonably short time the two halves of the plate were fused together again. As soon as the job was done I thought of the meat on the Rover roof and lost no time in "hitting the trail" for Liebig Bore, where I planned to empty lots of basins of water over myself before starting the last leg of the trip.

By mid-afternoon the outline of the big fan on top of the windmill tower at the bore began to take shape, and the big black stock tank at its base presented itself soon afterwards. A very slight air movement was causing the blades to revolve lazily, but with such a good-sized tank usually brimming with excellent quality water from the sub-artesian supply, the long

Top and Bottom: The tragedy at Liebig Bore. The bodies of dead and dying stock lay everywhere

trough would be full for the mobs of cattle and horses which congregated there. The ball valve assured that the need for supervision—which entailed making the long trip from Papunya—was kept to a minimum. As was the case with our bores at Giles, also close to sizeable mountain ranges, the underground supply was prolific.

Visions of basinfuls of cold water kept coming into my mind's eye as I stewed in the vehicle with a combination of sweat and fine dust covering seats and steering wheel. The shimmering expanse of bare open dirt was finally all that separated me from the airy bathroom, and the mirage at this distance made the whole scene appear to be floating in mid-air a metre off the ground. Soon the junction of tank and mill became clearer, and it was only then I began to sense that something was very wrong.

I had expected to see the stock jostling for position on either side of the strong trough in the furnace-like heat of the flat, but there was not even one animal standing near it. Dozens of bullocks lay about in the dust, and several horses with heads drooping to the bare ground stood under an almost dry mulga tree 100 metres away. As I drew closer I noticed the awkward way the bullocks seemed to lying: legs up in the air, faces buried in the dust. Their bloated appearance immediately told its own story, and in a minute I was out of the Rover and gazing about at the scene of utter devastation.

Dead and dying cattle lay everywhere, but why? I looked into the trough, and saw that it held only a few dry sticks and a deep layer of dust. Then I climbed on to the outlet pipe and looked over the searingly hot rim at the emptiness on the inside of the tank. A few centimetres of green slime covered the bottom below the pipe, and although the fan blades were turning, not a drop of water was coming from the mill. I couldn't believe my eyes. I remembered that when we had joined our road from Sandy Blight Junction to Liebig Bore, this tank had been overflowing with clear water, and we had even stopped the spinning mill-blades with the "pull out" gear to save their unnecessary action.

Making for the actual bore situated in the centre of the square formed by the legs of the tower, I found what had happened to cause the death and suffering of so many animals. The connecting rod joining the axle to the pump had broken at the threads where the big brass nut was located at the head of the bore casing so that, as the fan revolved, the loose wooden stem with the stump of threaded bar attached rose and fell in mid-air. The vital pump rod remained stationary, but by a miracle it had been prevented from dropping down the hole and out of sight because the huge nut had become caught on the lip of the steel bore-casing. As a result the water remained right where it was, deep down in the earth, with the motionless pump valves immersed in it yet unable to transport it up and into the tank.

With the initial feeling of stark catastrophe all about me still implanted on my brain, I roused myself into the action which was obviously needed as quickly as possible. After a study of the broken union, I noticed that a little over a centimetre of thread remained on the swinging stub, and that the nut was quite a thick one. If I could anchor the rods from the bowels of the earth to prevent them from slipping down, and raise the whole column to meet the loose end from the top of the mill tower above it, then I might be able to use the nut to join them together. Half the thickness might hold the lower piece and half the upper to serve as a temporary union, but taking into consideration the great weight of the long string of rods, would only half a nut be strong enough? I had only limited tools with me on the Rover. It was possible that when the strain was applied the threads would simply shear off, allowing the lower string of rods to drop away out of reach forever. But one more glance around at the surrounding ruin was sufficient to convince me that I should try.

The first thing to be done was to prevent any possibility of that nut slipping off the steel rim. With infinite care I managed to encircle the rod under the nut with as many turns of heavy fencing wire as I dared, before forming a loop with a figure-eight knot. My Land Rover hydraulic bumper jack was just the shape for this operation, and in a matter of moments I had placed it firmly on the steel jack-plate with the adjustable cast-steel tongue engaged in the loop. As I slowly worked the handle, the nut lifted above the rim and for an agonising instant swung clear into the centre of the circle of casing. But the old bush standby of fencing wire held, and as I levered, the rod gradually emerged up from the hole until the nut was half-a-dozen centimetres above the casing. Luckily I carried an apparatus consisting of two parallel steel bars bolted together for use in clamping broken spring assemblies. With one bar either side of the precariously-hanging rod and resting on the rim, I screwed the clamping nuts on to the heavy bolts until they closed together around the rod like a vice. It was only a friction grip, but I put on the greatest strain possible short of stripping the threads, and tentatively relaxed the lifting jack. Once again it held firmly and I could breathe more easily at last.

The first mishap occurred as I began to operate the "pull out" gear to stop the fan turning. The "pull out" gear consists of a ratchet and cog winch which, joined by a light cable to the sail, moves the blades at the correct angle to catch the wind. If this vane is moved to a different direction from the plane of the fan, the breeze cannot catch the blades and the whole action stops, even in quite a strong wind. As I wound the handle the cog assembly snapped, rendering it useless, so I took a length of rope from the bundle of frozen meat on the roof of the Rover and climbed up the narrow ladder to effect repairs. Once I reached the apex of the tower I clung there precariously, endeavouring to grab at a blade as it rushed past. From the ground the revolving appeared quite lazy, but now that I was within centimetres of it, things looked very different.

Eventually, by working around the top clear of the knife edges slashing through the air, I was able manually to close the angle of the sail and so stop that potential guillotine from moving. Once stationary it was a simple matter to rope it in such a way that the hanging wooden pole was at its lowest point. Back I went down the ladder, and with the combined aid of the jack and holding blocks, I finally brought the two stub ends together. Now, with the use of my largest shifting spanner, I was ready to turn the nut. Hoping that the threads would mesh, I scratched on the stump of the rod a line half the thickness of the nut so that if it was screwed upwards to that point, then both ends would supposedly be held equally.

Luckily all this went as it should, and I was now ready to carry out the final crucial test.

After I had unwound the fencing wire from the rod and removed the clamp I was elated to find that the join remained intact, so it was back up the ladder to untie the holding rope. With the fan once more turning, the whole string of rods began to go up and down. This meant that the flap valves at the deep end in the pump were doing their job. The water would once again begin to be elevated and would finally spill out of the feed pipe into the tank.

The heat was terrific out there in the open, and the thought of the water made me aware how parched I had become myself. It was only then that I made the rounds of the tortured cattle which weren't yet dead. Some, completely unable to move at my approach, were breathing heavily with their enlarged nostrils half buried in the dust. Some were able to make a feeble attempt to protest at my proximity and others just rolled their dilated eyes in agony. I counted forty dead already, and dozens were in the last stages of dying; but now, with the water trickling out of the inlet pipe, an attempt could at least be made to revive some of them. I realised that before the water level in the tank rose enough to reach the outlet pipe to the trough, many more of the animals would be dead, so I set to work immediately. After I had splashed handfuls of water into their mouths some of the stricken beasts rallied enough to drink the life-giving liquid from a tin basin pushed under their muzzles, and I had the satisfaction of seeing several actually struggle to a standing position. Others were quite obviously too far gone to try, and after an hour of carrying basinfuls of water from the small stream out of the pipe to the prostrate array of bodies scattered about I obtained a fair impression of their chances for survival.

Revolver in hand, I returned to four of the more unfortunate steers and terminated their suffering. One big bullock in particular wasn't responding, but I decided to persevere on into the early evening. The brutal sun was at last beginning to dip in the brassy sky and there was a hope that air currents resulting from the change in temperature might cause the windmill to pump at a faster speed. The thin trickle would then develop into a fuller steady stream, and at last the dust-coated trough would fill. It was ironical that the wind necessary to activate the watering system had also been the cause of the breakdown, for I concluded that this had taken place at the height of the gale-force winds battering the mill two days before. The supersonic speeds to which the fan had been whipped up had obviously proved too much for the rods.

The next step, before taking a shovel to clean out the debris and sand from the choked-up trough in preparation for the water which should soon be flowing in, was to call up the outside world on my transceiver and have a message relayed to Papunya. I had to advise them of the situation at their Liebig Bore and tell them that my repair attempt on the mill was only temporary. Although I hoped it would last to cope with the immediate crisis, a more permanent fix would be needed to forestall any repetition of the calamity. First I would have to send a message to my base communication centre at Woomera, and Ron could then repeat it all straight away to the Alice Springs Flying Doctor base. Once in Alice Springs, the emergency system would go into operation and Papunya would know about it within the hour. I contemplated this wonderful system and the marvels of progress as I transmitted my call sign to old Ron.

Although I was becoming dog-tired from the previous week's activities, there was little sleep for me that night, camped as I was beside the windmill. When I wasn't making periodical visits to the helpless bullocks with my basins of water, I couldn't keep my eyes from the rotating fan. In the early hours of the morning I was elated to feel a strengthening of the air movements, and soon, sure enough, the blades of the big fan began to blur together as they gathered speed. The resulting flow half-filled the inlet pipe and the water level began to rise. It had almost a metre to go before the trough would begin to fill, but as I studied the assortment of pipes and joints I devised a way of reducing that depth by half. By screwing the outlet goose-neck through an arc of about a right angle, I could make the nozzle half a metre lower than it was in its upright position, and with the aid of a sheet of loose iron the flow could be directed back to the trough.

So it was that after I had worked throughout the night, the trough actually began to fill, and by the time the first streaks of daylight appeared, silhouetting the stately peak of Mount Liebig, the water had risen to almost twenty centimetres. I dragged some of the younger calves bodily to the trough and pushed their muzzles into the cool water. One or two couldn't drink at all, not having even a spark of energy left in their weak little frames, but others did. Eventually, sensing the activity, the drooping horses plodded over to join in. I kept well away from the magnetic attraction of this long wet trough and was fully gratified to see a succession of animals gathering, their agony at an end.

When I revisited the big bullock with which I had persevered for half the night, I found him to be dead. I could only hope that I had brought him at least some measure of relief in his last few hours of life.

My task here was done. After I had at last emptied those lots of basins of water over my grimy, sweaty body, I replaced the rope on the load of meat and climbed into the driver's seat of my faithful Rover. It looked like being a good day, I thought, as I watched the stock jostling for position at the trough, the spinning fan on the windmill, and, more important, the pump rods rising and falling at every turn.

Transferring my gaze to the now illuminated foothills of Mount Liebig and the majestic bulk of it rearing up into the burnished morning sky, I had a multitude of mixed thoughts as I slowly drove away and on to the start of my road out west.

Chapter Four

SANDY BLIGHT JUNCTION

As always, when events of a disastrous nature have happened, however calamitous they may seem at the time, I have always seen some good come of them. If I took the trouble to search for a benefit brought about directly by some unforeseen accident, then I would always find one. So it had been in this case, I thought. Who would have believed that any good could possibly come from our ration truck exploding in the desert? Yet on this catastrophe had depended the fate of a mob of suffering animals, to say nothing of the cessation of Mrs Snook's agony.

During the trip to Sandy Blight Junction, memories of making the road on which I was now travelling came floating back.

We had started work on this road at the beginning of 1960. It had originated from our Gunbarrel Highway near Giles, and wound in a northerly direction for over 400 kilometres around mountain ranges, salt lakes, and sandhill belts. At the time of my original survey there was nothing there but a remote expanse of spinifex north of a huge mountain outcrop with a 920-metre high peak named Mount Leisler. A careful reconnaissance to the east and west had resulted in the exact location of the junction, which I later fixed with a latitude and longitude observation to the stars after the actual road intersection was made.

A couple of weeks before this star observation I had been obliged to make the long trip to Adelaide. It had been only a quick visit to have some large-scale replacements made to my Rover, which had suffered extremes of hard use since the beginning of the year; and I also had several items of personal business to carry out for various members of the team. In the meantime the others in the party would stay at the camp to prepare the heavy machinery for the next onslaught into the bush.

Just before starting the journey back from Adelaide, I had the nagging sensation that my eyes had grit behind the lids, and no amount of rinsing would relieve the feeling. Bushmen call the condition "sandy blight" because the eyeballs feel as though they are coated with sand. With every blink the lids grate over them, and the inflammation makes the eyes themselves extremely painful, glueing them shut after each night's attempted sleep. By the time I reached Woomera I could hardly see to drive, and so as soon as I could I sought the assistance of the doctor there. I had planned to stay in Woomera for a day, and during this time I made several visits to the hospital for treatment with eye ointment, which gave considerable relief. The doctor urged me to stay until my condition eased, but I couldn't help thinking of the boys back in the bush. By now they would have finished tuning up the equipment, and they would probably be ready and willing to shift camp. I also wanted to get back to continue with our huge project, so I planned to creep away at first light on the following morning before anyone was aware of it.

This wasn't so easily done, for I had been temporarily accommodated in the same house as my very old friend Padre Stewart Calder. This man had then been living in all conditions at Woomera for about fourteen years, almost since we started the whole project. He had been a tower of strength, tending the needs of the floating population in every conceivable way, and a bundle of constant energy, keeping his finger on the nerve centres of everyone about him. He knew very well what my intentions were and had a doctor call that night with a supply of eye balm and a further lecture aimed at delaying my departure.

Amazingly my eye condition seemed to improve during the evening, or so I tried to convince the good padre, in such a miraculous fashion as to enable me to head off at first light. However, the blight worsened with every kilometre, and was now aggravated by an increasingly sick sensation, so I camped alone in the mulga only 250 kilometres on my way. As I emerged into the Coober Pedy opal fields environment I could scarcely see the wheelruts winding through the saltbush prairies, and the nausea in my stomach made any thought of food repulsive.

Next morning it took half an hour to soak my welded eyelids open, but after a liberal application of chemical I managed to carry on without stopping except for fuel and running repairs to my eyes, until eventually I reached Victory Downs station. This had been the start of the first road we made across Central Australia, the 1500 kilometre stretch later named the Gunbarrel Highway, so from here on any complaints I had would have to be directed towards myself.

Meanwhile the thoughtful padre had consulted the doctor, who had in turn radioed ahead a message which was waiting for me at Colin Moreton's homestead. The gist of the signal indicated their concern and asked that I remain where I was for observation, possibly by the flying doctor. After my ultimate escape, I found a further radioed message waiting at Mulga Park, my next port of call. This time it had been addressed to Dave and Ted Fogarty, asking them to virtually hogtie me until an examination could be made. As I sat groaning in their comfortable station house, I began to understand just what it was like to be a bushranger on the run. However, after an enforced rest which only a little conviction on their part convinced me was necessary, I repeated the act whereby I assured my captors that the spell had brought about a marvellous improvement—so good, in fact, as to allow me to depart restored to health once more almost completely.

By the time I reached Giles again, by means of sundry eye treatments and much clutching of my ailing mid-section, I felt as if I'd been through a concrete mixer, and the gritty feeling in my eyes had continued unabated. Flies were often the initial cause for this condition, which was often found in bushmen and Aborigines, and I felt that dust and sun glare didn't help the situation either. The ointment had less and less effect but it was good while it lasted.

The wily padre and doctor from Woomera had of course a direct radio contact with our base transmitter at Giles weather station, and it was all I could do to convince them that my current plight was not worth a flying visit by an expert. Before anyone could start thinking of other ways to delay my return to my waiting camp, now only 450 kilometres away, I refuelled and departed. There would be time to make a recovery once we were under way with the new road.

At last, after over a week of eluding the radio waves reaching out for me as I laboriously strugged from station to station, I was on my own with no fear of further signals other than from my own portable set in the Rover, which could very easily be left closed in its case. I must admit that I was pleased at long last to reach my camp. I greeted the party with bleary eyes and one hand pressed firmly above my belt buckle.

There were distant sand-ridges to the east and west of my proposed junction, but after much reconnaissance I found I could complete the intersection with a long straight east-west length of road which would clear the nearest obstructions. As it happened, this feature eventually proved helpful to future users, who widened the finished road into a broad clearing to make a quite useful airstrip. I had marked the actual point of intersection with a deep set of wheeltracks, and when I rejoined the boys, we turned the bulldozer in that direction. Our earlier intention was to make a further 150-kilometre link northwards and eastwards to join up with the existing Tanami–Halls Creek track at Mount Doreen, but as it happened the 1000-odd kilometre access to Marble Bar became more urgently needed. Fortunately we had progressed only twenty kilometres towards Mount Doreen, so after the change of plan we didn't have to backtrack too far. The remaining country in this direction was reasonably open, and following surveyors would have no difficulty in travelling over the area without a road.

When the bulldozer arrived back at the proposed junction site, Doug first manoeuvred the machine to point in a westerly direction, away from the Liebig Bore. This was done to form the actual junction and also to give us a start of several hundred metres of made road when the time came to continue. While I guided the direction in my Land Rover, the heavy blade of the bulldozer gently skimmed the ground surface, picking up a load of spinifex. We then turned around for the return journey. As he regained the corner Doug widened the first cut by double the length of the blade, then carried on the straight line in an easterly direction towards Liebig Bore. The others could now follow in the wake of the bulldozer until a distance of about two kilometres from the junction had been reached, and here they would make our overnight campsite. While the rest of the party settled down for the evening, the new section of road was widened and graded.

I was pleased to notice in my diary that it was now Saturday: this meant, even out here, that a day's spell must be taken. As we were to camp so near to this newly-formed intersection, I decided to make use of this opportunity to carry out the astronomical observations that would fix its position accurately for all time. The latitude and longitude thus obtained would also help me to determine the course of the new roads in either direction. I couldn't waste the chance, sore eyes or not.

That particular night of star readings stands out above all others in my recollections of physical miseries endured, although the stars themselves shone brilliantly, unobstructed in all directions in a clear moonless sky. Five hours and a half a tube of eye ointment after commencing, I had obtained enough times and angles to produce the nearest possible fix available in the field. An average of two or three stars could be sighted before the eye condition took over and made focusing on them and the optical micrometres impossible. Another application of the treatment lasted a further few readings.

Somehow by midnight I had finished and laboriously packed the instruments away, carefully placing the valuable set of observations in the built-in box in the Rover. I fully intended to complete the calculations on Sunday when we were stationary in a camp for once, but doubted if it would be possible. For some reason the words "sandy blight" kept repeating themselves in my mind, and it was during the worst period that the name "Sandy Blight Junction" loomed up as a name for this crossroad. Apart from the inflamed red eyes, I'd been plagued by the sand-ridges which had governed its geographical position. In the morning I informed the camp of the name of our new corner.

By the time we shifted camp Quinny had continued on to Giles for a load of diesel, water, and rations but had as yet not returned. We were in need of the supplies, but having stumbled upon such an appropriate name for our newly-constructed junction I was anxious to concentrate on the calculations and make an aluminium signplate to label it for all to see. Quinny and his whereabouts could be dealt with later.

All during that Sunday, as the sums progressed, I battled with the hoards of flies, interspersing these struggles with so many applications of the eye ointment that I used up most of what was left. At last, late in the afternoon, I proudly drew a line under the final values: latitude $23°11'58''S$ and longitude $129°33'35''E$. One of the aluminium plates carried for the purpose was soon out on the steel jack-plate, and with my set of alphabet punches I stamped out the information under the bold heading "Sandy Blight Junction." Some years later I began to notice that name labelled on maps. Every time I see it, memories of that torturous night still materialise in my mind.

With the signplate ready at last, the next problem was to find a suitable tree to use as a supporting post. I had in mind a solid monument which would last a long time, but as there weren't any logs thick enough near the camp I postponed that part of the operation and turned my attention to a job which badly needed doing: that of cutting everyone's hair. As soon as that was over, as if the tasks would never end, I rigged the aerial for a radio contact with Giles to ascertain if Quinny had left with the supplies.

He had already been there and departed again for our camp two days before, and as it was only a 400-kilometre journey, he should have been back by this time. Not wanting to delay our Liebig road any further, I refuelled my Land Rover and headed back to Sandy Blight Junction and south on our Giles road to find him.

After sixty-five kilometres I was relieved to see his stationary truck looming up on a bend we had made around the terminal end of a long sand-ridge next to the Davenport Hills. Frank Quinn was sound asleep on the 1400-litre water-tank where he lived, but as the noise of my approach grew louder I noticed his brown shiny head appear out of his swag roll. It didn't take long to discover that his truck would no longer move without a great deal of help. As he had ground the heavy vehicle with its huge load around the sandy bend, a loud clatter had come from underneath and an instant later the truck stopped dead. The transfer gear-case had shattered almost in half, and the cogs inside did the rest. Having plenty of water and food, Frank had merely settled down to wait for whoever would eventually come to him, as he knew they would. Accidents like this had happened many times in the past.

Still not wanting to delay making the new road, I loaded our sealed box of fresh meat on to the roof of my Rover, turned around, and headed back for camp there and then. In the morning we would send the grader back for Quinny to tow his truck-turned-trailer indefinitely until a new transfer gear-case could be sent.

In the early hours of the morning I rounded the corner at Sandy Blight, knowing that there were only two kilometres left to go, and that I could idly sleep for what little was left of the night. The news of Quinny's plight sent Scotty back on his grader first thing to begin the towing while Doug and I continued on with the road. The grading could easily catch up when Scotty returned. At least we were on with the job once more.

Late that night the grader with the crippled supply truck in tow chugged into our camp, now another fifteen kilometres further on, and I again contacted our base transmitter to order up a replacement for the broken transfer box. All went well until the sixty-kilometre point, when the next calamity occurred.

There had been no response to my flashing mirror guiding the direction in which the bulldozer should come, so I drove back to find out what the hold-up was. Doug was still in the operator's seat, but to my horror I discovered that he had been badly injured. Apparently the spring on the blade-lift lever at the rear of the dozer had broken and the handle had snapped back, almost shattering his hand. Immediately realising that there would have to be a change in proceedings, I pulled shut the throttle and dropped the compression lever to stop the big diesel motor. The steel handle, no longer held in place by the spring, lay loosely. I soon had Doug in the Rover and back to camp. Examination revealed that nothing seemed to be broken, but the hand was severely bruised and swollen to twice the usual size. Although it was giving him much pain, he didn't think a rush radio call or long trip back was indicated so we decided to rest it up for a while and see if hot-water bathing would restore it to normal again. Scotty would take his place on the dozer while Eric did a crash course in grading.

This arrangement worked well until we reached the Ehrenberg Ranges, the halfway point between Sandy Blight and Liebig. By this time Doug could use his hand without discomfort, and as Rex had repaired the fault in the dozer before Scotty had taken over, we were all at last back to our usual jobs.

Eventually we received a radio message saying that the new transfer case had arrived in Giles by Bristol Freighter, so we made the trip there in Rex's truck to collect this vital piece of equipment and some more supplies. By the time we had progressed to a point 150 kilometres from Sandy Blight, Quinny's supply truck had been restored to as good as new by Rex and helpers. We were now in an area so densely wooded that nothing larger than my Rover could venture off the bulldozed cut through the mulga scrub. Conditions had been heavy going for days, so while the replacement of the newly-arrived transfer box was going on I decided to carry out a reconnaissance survey to plan the course of the rest of the road right into Liebig Bore.

That return trip of eighty kilometres through a concentrated mass of mulga scrub took a heavy toll on my Land Rover, especially the tyres, which became staked with monotonous regularity and needed mending at intervals throughout the whole day. I was grateful that my eye condition had eased at long last and I could work without the extreme difficulty it had caused; but in one way, now that it was over, I was glad it had occurred. After all, it had given the name to not only a major road junction but also to 380 kilometres of road which is in constant use to this day.

After each tyre changing and mending operation, I found that the Rover was becoming increasingly difficult to restart, and I was relieved when I gained the comparatively open area around Liebig Bore where I could work on the problem for the return beating back to camp. I discovered then that with the punishment it was taking the fuel pump was the culprit, so I replaced it from my spares.

A small group of Aborigines from Papunya had appeared silently while I was buried in the engine, and when I looked around, having finished, I received quite a start at the unexpected company. They had left the settlement for a sort of holiday, usually referred to as a walkabout, and I suspect they were as surprised to find a strange vehicle at the bore as I was to find them.

The result of that day's bush-bashing gave me the exact route from the head of the road for the last link-up to open up the way clear through to Alice Springs. Eventually, provided we were successful in completing the road on from Jupiter Well, a new north-west road would be opened up right across Australia to the Indian Ocean.

The link-up was completed on 16 September 1960 when, with an exaggerated flourish, the bladeful of dirt in front of the bulldozer was tapered off and the grader right behind smoothed it out on to the existing station track. The 180-kilometre length of road on which I was at present driving was finished.

I remembered that I had immediately made a pilgrimage into Alice Springs to obtain some items which we always seem to be needing, camping in the bush a few kilometres short of the town to shave and change my clothes. On the return to camp I had run into the most violent storm I had ever encountered in that country, with not only sheets of lightning and explosions of thunder but hailstones as well. The rain had turned the track into a river, and I could only keep on the compacted wheeltracks by driving along that part of the flooded expanse that I could see was free of trees. The deluge lasted all the way to Papunya, and the hail smashed on to the aluminium roof of the little Rover as it waded along, illuminated from time to time by brilliant flashes of lightning. The one act of opening and closing a station gate left me as soaking as the usually dusty mulga trees lining the road, with my hobnailed boots brim full of water.

I was glad this storm had not reached as far west as my new road, because this was as yet uncompacted, and it would have been hopeless to attempt to drive over it if it had also been under water. As the final straw, after leaving the raging elements behind, one of my tyres finally succumbed to an old mulga stake which found its way into the tube.

Back in the camp we had filled all our tanks with the good quality water from Liebig Bore before turning around to drive back over our own new road to Sandy Blight Junction. Once there we would begin the road to the north-west, taking off from the end of the short beginning we had already made.

All these multitudes of memories now made my long trip back to the camp pass by almost unnoticed. Then, returning abruptly to the present, I thought of all that unfortunate fresh meat on the roof of the Rover in this heat. It must be installed into our refrigerator box as soon as possible: even after my unscheduled delay at the broken-down windmill there was a chance it was still usable. We would soon find out.

Chapter Five

MOUNT LEISLER

Just short of the Sandy Blight Junction, the stump of the thick tree I had chopped down for use as the signpost showed itself alongside the road. I was satisfied it had been the best one available on the whole stretch of road. After chopping off a length of three metres I had barked it, discovering only then that it was impossible to move manually. It was a terrific weight. While waiting for the boys to rejoin me on our way back to launch off once more to the west, I shaped a squared section at one end with my razor-sharp survey axe. Here, slightly down from the top end, to allow room for a piece of sheet lead covering, I would nail on the aluminium plate with its information for future travellers. I always capped such signposts with lead in an effort to keep out the weather.

When the bulldozer arrived, we used it to move the post into a position where we could rope it on to the blade to carry it the rest of the way. A huge hole was dug at the junction and then, with the great post roped upright to the elevated dozer blade, it was manoeuvred into place and lowered gently into the excavation with the winch. After the hole was filled in and rammed with the crowbar Paul normally used as a stove flue support, an old staked grader tyre was looped over it, followed by a truck tyre and lastly one of my useless Land Rover tyres carried on specially for this purpose. It soon began to resemble a gigantic set of deck quoits. When the tyres had been filled with sand and the whole structure painted white, we had left behind yet another monument at Sandy Blight Junction.

Intent now on meeting up with my party as soon as possible, I carried on straight past the still new signpost. There wasn't any need to stop and see whether the bulldozer train had yet reached the corner, for it took not a great amount of tracking ability to notice if forty tonnes of equipment had happened to pass by.

The boys had come 160 kilometres since I left for Alice Springs, and this put them within only ten kilometres of Sandy Blight. So it was that no sooner had I passed the corner than I noticed a dust cloud ahead. A little later the din emitting from the mobile mass of steel as it clanged along came through the hot desert air to drown out the comparative quiet of my vehicle's approach. Five days and nights had passed since the ration-truck fire, and on the sixth day I was back, after having crammed in more adventures than an average person has thrust on them in a decade.

Quinny and Scotty had long since rejoined the camp with the much-needed rations from Giles, and after the boys had had a good feed the inelegant progress of the unwieldy train was able to resume. I drove off the road as we closed the distance between us and climbed stiffly out of the cabin to greet them. Doug, who was in the lead to set the pace, throttled down his big diesel engine, causing the bulky conglomeration behind him to clash together momentarily, then, throwing out his clutch lever, he stopped the whole affair dead in its tracks. The noise was deafening and I thought that the operators, as always on our work out here, really earned their keep.

A quick reassurance as to how things were going with the convoy and the boys in the group was all the time we wanted to hold up this slowest unit. After a long drink from Doug's "dust-turned-mud"-caked waterbag, I clambered down over the shiny Caterpillar tracks and he was off once more. All I could see of his features was the whites of his eyes and his pink tongue, showing in startling contrast to the mask of mingled sweat and dirt from the clouds of swirling bulldust accompanying the movement. The water from the bag now revealed sparkling white teeth as well.

The others in the party were by now right behind, so without further waste of time I drove my Rover with its load of meat alongside the little fridge trailer and untied the ropes. In anticipation of this, Paul and Rex had the compressor motor going and the inside of the little box was as cold as possible, even without its canvas canopy. We dumped the clumsy canvas bundle off the roof on to the road and freed the rest of the remaining ties. They hadn't been moved since I left Alice Springs and now, as we unfolded the camp sheet, we all wondered whether we were going to eat well that night or not. It would be deplorable to be forced to leave the load out on the plains for the dingoes and eagles to feast upon after all this.

As we opened the last of the coverings, pieces of still solid dry ice fell away, and we found that the meat was cold and in as good condition as when the butcher and I had rolled it all up on the ground in the Alice. In as much haste as we could muster we handed the pieces to Paul, who stacked it into the cold box while the others beat at the swarming flies with mulga branches. One way to help the temperament of hardworking men is to endeavour to keep good food up to them, and now the geniality of the team for the next week was assured. Not that our little party would have been any different either way, as they had been out in these conditions with me through thick and thin for many years, but I was always pleased to be able to make conditions better if at all possible — which so often, regrettably, it wasn't.

After the meat transfer had been accomplished, Quinny and I resumed our usual position in front of the train to prepare the dinner and overnight campfires in advance. We soon passed Doug and drove on to the road ahead once more, and I thought that everything was as it had been before the unscheduled disturbance, except that Paul now travelled with Scotty in the little workshop Rover.

When we arrived back at Sandy Blight Junction to wait for the others I realised that the right-angled corner, or almost any other corner for that matter, would be far too sharp for Doug's long string trailing out behind. Planning a scrub-free path in a long easy arc, I returned to guide him around considerably short of the actual intersection, and soon we were all settled on the Sandy Blight Junction Road, heading south at last, with 400 kilometres left to go.

Mount Strickland, a large outcrop on the northern end of the Kintore Range, gave us something to break the monotony of endless spinifex and sandhills. We had already been catching glimpses of it for a day, and now it loomed up reasonably close to the west of the road as we crawled past. Then, on the eastern side, there was the impressive outcrop of huge red rocks and loose boulders to which I had guided the road four months previously.

I had first found them on my original reconnaissance from the last camp when, leaving the major servicing of the plant to Rex, with the rest of the camp to help him, I had gone off to plan the next day's route. Keeping clear of the stony foothills of the Kintore Range, I had noticed this outcrop across a large spinifex plain in the direction I was heading. I slowly crossed the distance over the hummocks of growth and sand, and long before I stopped among the exciting conglomeration of smooth rounded rocks, I knew I would be making the road as close to them as possible. Future travellers couldn't fail to be impressed by the spectacle, for some of the rocks were larger than houses, spherical, and with interesting-looking crevasses among them. As I opened the door of the Rover a movement caught my eye at the entrance to one darkened cavity, and a magnificent specimen of a dingo which had been watching my approach turned slowly and retreated into the blackness. Now that I had decided upon the course the next dozen kilometres of road would take, my work that day was finished. All that had to be done now was to plot the return journey to camp by the straightest route possible, clearing any rocky obstructions on the way.

When I reached the camp I was full of the story of my discovery, and as it was the last day of June and bitterly cold, we sat around our roaring campfire till late in the evening while I held forth on the great excitement in store for everybody on the next day. What a sight they were in for, I told them; and in addition it was a certainty that this wondrous phenomenon of nature had never before been gazed upon by white man. How privileged we were all going to be! Such peaks of excitement as this, coming our way so unexpectedly from time to time, really gave us the enthusiasm to carry on. At length I came to the end of my spiel, quite exhausted, as I'm positive was my audience also, and Doug sleepily remarked to the camp that he supposed they were only a few stones around some goanna holes. So often, when months of sheer hard work had begun to nudge slightly at the men's endurance, I would return to camp from a forward expedition bringing glowing tales of what lay ahead, attempting by this

means to revive their wilting spirits. Not that my stories were without basis, but perhaps sometimes I might be guilty of overpraising the particular feature which I hoped would serve as a goal to tide us over an especially arduous period. By now they all seemed to have become used to these glamorised descriptions and I felt my sensational news was falling on partially deaf ears.

In this case, however, I wasn't at all to be put off, and I stepped up my excited harangue, likening the scene to one of the wonders of the world. My exaggerated eagerness served only to convince the weary group around the fire that I was raving, but at the same time I was complacent in the knowledge of what I had perceived with my own eyes. The boys were in fact certain to be in for a pleasant surprise.

Bright and early the following day, the ear-splitting explosions of the unmuffled exhausts from the pilot motors of the bulldozer and grader disturbed the normal stillness of the crisp morning air. As soon as the main diesels had fired, Doug and I were on with the job. As I guided the first cut of the bulldozer with a flash of reflected sunlight glinting off my worn rear-vision mirror, I noticed that Scotty with his grader was hot on the heels of the Caterpillar tracks, on the off chance that there might be something to see after all. There were some scrubby trees to break through before the open expanses of spinifex were revealed ahead of us, but my pile of rounded rocks was still too far off to be seen clearly.

Eventually the distance lessened until I was able to drive to a spot alongside this amazing natural phenomenon. I flashed the mirror at the hazy bulk of the waiting dozer, and as it came towards me I sat on the roof of my Rover with as smug a look as I could muster. At last Doug throttled down the motor, with the great steel blade almost touching the tailboard of my comparatively frail little vehicle, and climbed down over the grouser plates to jump heavily on to the ground with a broad smile on his face. I joined him, and in answer to my silent look of query he wholeheartedly conceded that here was something which for once actually lived up to my effervescent sermon of the night before. I suppressed the urge to say "I told you so," and together we walked over to examine it properly. Clambering around, between, and through the various formations, we were soon joined by Scotty, who reluctantly agreed with Doug about the grandeur of the spectacle. We scared another dingo and after a thorough examination left the place with a feeling of restored faith in working out here in the deserts of Central Australia.

We continued with our roadmaking until a little further on we reached a good camping area with more dry wood available for our fires on these freezing nights. Then we turned as usual to widen out the track. When the rest of the camp caught up with us Rex and Eric turned out to be suitably impressed with the historic day's events, but I was afraid it would take more than this to cause old Paul and Quinny to turn cartwheels, although they didn't argue about it looking all right—as far as rock outcrops go.

As we trundled past the monoliths now with that self-same bulldozer I wondered if they would score a second glance from the party this time. The boys had been living all the year out here, and this particular discovery was still too fresh in their minds. It would take a longer spell away from it all in another section of the bush to give them time to reflect upon it, and in years to come they would perhaps look upon it through different eyes.

The next great point of interest for our train to pass by was Mount Leisler, with the historic explorer's tree at its base. Mount Leisler, a most spectacular peak rising 900 metres above sea level, rears up out of the southern end of the Kintore Range and towers over the entire countryside. It is the highest mountain in all that part of north-west Australia right across the rest of the 1000 kilometres to the Indian Ocean. Nothing comes up to it for the same distance north to the Timor Sea, and several peaks in the Petermann Ranges 200 kilometres to the south are the only ones topping it all the way to the Great Australian Bight, a distance of another 1000 kilometres. Mount Liebig and others in the Haast Bluff and MacDonnell Ranges area take over 200 kilometres away to the east, but nevertheless Mount Leisler is an outcrop to be reckoned with. For us, keeping in mind the overall survey of Australia for which all this access of road network was intended, it was an obvious "must" for a future trig station hill.

My first sighting of it had been at a distance of eighty kilometres, while I was making a reconnaissance trip to plan the route of the road past the Bonython Range area south of Lake Macdonald. The road would have to bypass the 200 square kilometres of this huge dry saltpan, and to make matters worse it usually happens that the nearer they are to salt lakes, the higher and more concentrated the sandhills become. The new road would have to clear these, plus the rocky jumble of hills comprising the Bonython Range; and so for a better sighting of the terrain ahead I had decided to climb the highest of the latter group.

The Land Rover couldn't ascend the last section of the boulder-covered hill, so I had to make the rest of the distance on foot. As I scrambled to the top I saw in the north-east, to my great excitement, the striking shape of the feature which, according to plotted bearings and my prismatic compass, must be Mount Leisler. My first impression was that it resembled pictures I had seen showing great jagged icebergs rearing up out of the sea: because it was some distance away, it was even coloured a similar blue. From its western end it rose steeply to reach a sizeable peak before dropping away suddenly, leaving a cliff-like face on its eastern edge. Then and there I determined that my road would pass as close to this spot as was practicable, not only for the convenience of later surveyors, who would need easy access for their trig work, but also to bring such majestic scenery within easy reach of any future travellers.

From the time of that first sighting it took six weeks of hard work to actually reach Mount Leisler; but each time I have used that road in the succeeding fifteen years to date, I have never regretted the effort. It was another occasion on which I returned to my camp overflowing with eloquent descriptions of a sensational discovery which lay ahead of us, just awaiting our arrival. Once again I refused to be dampened by the usual jocular indifference caused by my news, and once again I was complacent in the knowledge of the change in attitude that would result when the party eventually saw it for themselves. I remembered asking them if they had ever heard of Mount Everest, and after the lazy affirmative answer, I carried on by conceding that although this wasn't quite as high, it was really something in this otherwise relatively featureless country.

After seeing Mount Leisler for the first time, I noticed some more smaller but nonetheless sizeable hills slightly to the east. As they were only about twenty-five kilometres distant, and another positive trig site, I determined to bring the road past them as well. It was alongside these Davenport Hills that Quinny had later broken up his transfer gear-box.

Mount Leisler of course could now be seen constantly by our

cavalcade as we drove ponderously towards it in the cloud of dust stirred up by the sheer weight and bulk of our train. I was sure the eyes of even my blasé team would wander to study the peak again. In the interplay of light and shadow caused by the sun's rays, it looked different every time we saw it.

When we had first reached the foothills of the mountain with the new road from the south, it was near the end of June and into the bitterly cold time of the year, with the temperature at night sometimes down to freezing. The Sunday before closing the distance to the foothills, we had decided to use our day's rest for a picnic. We drove in our two Land Rovers along my original survey wheeltracks, which would be transformed into a road in the following few days. All of us were there except Doug and old Paul, who were agreeably impressed by the towering summit now only twenty kilometres ahead, but not so much that they could be urged to put in the day bush-bashing towards it for a premature closer look. They put some food into an empty tomato-sauce carton for our dinner, but were themselves content to lie in the sun on their swags reading Westerns.

When we had arrived at the stony lower slopes of the mountain, we carefully retraced the wheeltracks which I had made when circumnavigating it. The topmost peak beckoned us with magnetic attraction. Leaving the Rovers as close as possible to the western base, Quinny and I set off for the eastern point of the cliff that formed the highest point, while Scotty, Eric, and Rex attacked it from another angle up an interesting-looking crevasse. Our parting words were that we would return to camp separately if and when we regained the vehicles. The tracks now had been defined by the passage of three sets of wheels over them and were easy to follow.

Every time we thought we had made it to the top, the crest above us proved to be false, with another soaring up ahead. This is a very usual finding with anybody who climbs hills, and the effect is useful in that it serves to break the whole effort into a series of smaller steps, so that the curiosity of the climber is retained. In all situations anywhere, if a project of any magnitude whatever can be broken up into smaller parts, the whole thing will be achieved before the participants realise.

As we climbed higher, the distance of the visible horizon lengthened. We could see Lake Macdonald to the west, and then the vast white salt expanse of Lake Mackay began to reveal itself 100 kilometres to the north-west. Lake Mackay is 100 kilometres long and covers an area of about 5000 square kilometres with a sheet of dry salt. It was between these two treacherous boggy stretches of salt-coated blue mud that I had decided to build my road north-west to the Indian Ocean. There is a seventy-kilometre strip of hard but sand-ridge strewn ground separating them. Now, of course, our road is there, and so, due south of Lake Mackay, is our burnt-out ration truck.

At long last, two hours after leaving the Rovers at the base of the climb, we approached the summit of Mount Leisler. The desert now rolled away to the flat skyline for almost the full circle, and as we sat down on the rocks to gaze on the scene no words for a time passed between us. It was a sight seen by so few white men, and we reflected on the nomadic Aborigines who, over countless thousands of years, had probably been right where we were at the moment, intent on finding food. A strong wind was evident at this altitude, and as the body heat caused by the exercise left us, it became suddenly quite cold, but we wanted to get a picture which would remain imprinted in our mind's eye forever before we started the descent. Occasions such as this made us all feel proud to be involved in this work which would benefit not only our Woomera project, which was its initial justification, but also the overall mapping programme of Australia.

The survey parties from the National Mapping Council were able to climb this mountain with their instruments in much faster time than Quinny and I did, and the enormous well-proportioned rock cairn they built at the top for the trig beacon will last for generations.

William Henry Tietkens named Mount Leisler in 1889 after a Mr Louis Leisler of Glasgow who provided the funds to finance his explorations into these regions, and he named the main Kintore Range after the new colonial governor Lord Kintore, in honour of the first year of his appointment. Tietkens would have been the first white man to have climbed to the point where Quinny and I were at present sitting, and I was certain that this "noble bluff," as the explorer described it, would have been visited by no white man since then. As we were to discover, the whole desolate region was absolutely waterless. William Tietkens was second in command of the Ernest Giles expedition in 1873–74 on which the fourth member, Alfred Gibson (after whom the Gibson Desert was named), lost his life.

In a journal compiled from a reproduction of his diary, Tietkens described Mount Leisler as a beacon warning all travellers to avoid its dry and waterless surroundings. When I perused his journal before the extensions of our work into this section of the country, I found one remark to be of particular interest. He stated that at the foot of this most easterly bluff was a bloodwood tree which he had marked by a T above the figures 5.89. He had made the ascent of Mount Leisler on Monday, 27 May 1889. An obvious course of action at once presented itself to me. Firstly, we would make all possible endeavours to locate the tree when we came to this area, and if we were successful I would surely make the new road lead to it so all future travellers could share in the historic atmosphere — providing, of course, that it happened to be situated in a place which would make the project topographically possible.

A bloodwood tree generally stands out from the usual collection of stunted mulgas, so it shouldn't be too difficult to find, especially as Tietkins had narrowed the area down accurately in his description. Also it would have to be of a reasonable size to allow room on its trunk for a marked blaze. I hoped that the tree would be unharmed by the ravages of time. What a wonderful thing it would be if all this could actually happen!

Chapter Six

TIETKENS' TREE

When at last we began to close the distance between the head of the road and the mountain which I had first observed a month and a half before, I planned a day-long trip to work out the details of where to guide the bulldozer.

An open spinifex plain lay between the Davenport Hills and Mount Leisler, and I travelled over this for about twenty kilometres before I saw several low, rocky hills followed by a small belt of sand-ridges. Then for a kilometre or so the way became clear to the actual foothills of the main bluff, only sparsely covered with occasional trees and scrub. Quite close to the foothills, where the rocky ground began its sweep up to become vertical near the top, the scrub thickened—a result of the water which cascaded down the slopes after the showers which deigned very occasionally to fall on this parched, forsaken region. Several hundred metres further on, open, stony flats took over. The whole area was dominated by the cliff towering up to 500 metres above the plain, and altogether it was a very pleasant locality.

On one particularly open area under the eastern shadow of the peak stood a bloodwood tree, bordered on its western side by a dry waterway along which a heavy patch of growth formed what was almost a hedge. Although not a wonderfully healthy specimen, with some dry branches attached, it was noticeable because it appeared different from the rest of the immediate scrub.

With great excitement in the anticipation of the imminent discovery, I approached this tree carefully, so as not to destroy any possible evidence of Tietkens' presence seventy years before, and climbed out of the Rover. On the southern face of the tree a partially grown-over blaze was plainly visible; and on the original trunk, still exposed and in perfect condition, was a cleanly-chiselled number 5. The words from Tietkens' journal came back to me—a T above the numbers 5.89. There was absolutely no doubt whatever that the T and the 89 would lie beneath the healed-over blaze, covered by several centimetres of solid wood and bark. Since the time when William Henry Tietkens had stood at this very tree with his wood chisel and hammer in hand, carefully cutting in his initial and the date, seventy years' gradual growth had been taking place.

The new road could easily be directed this way before progressing onward, and as no more reconnaissance was needed that day, I turned again for the camp.

It was with mixed feelings that, four days later, I guided the bulldozer to the tree. I wondered what the earlier explorer could possibly have thought about all this. In his day, motors in general were yet to be invented, but here we were carving a road past this particular tree, singled out from countless others, in order that all sorts of people could come in their vehicles and carry out their assortment of reasons for being here. The survey parties from the National Mapping Council would lead the field of subsequent users, the first of hundreds I have known who have since made the journey, all of whom must have stopped to examine Tietkens' marked tree.

So that we need not damage or mutilate the immediate surroundings any more than absolutely necessary we passed by the tree and continued on for a kilometre before turning the machine around to widen out the cut. After a fifteen-tonne machine has screwed itself around, the area is left pulverised, and the resulting devastation was not what we wanted in any proximity to this landmark. On the way back, Doug had climbed down from his idling machine and plodded over to view it, after which he grudgingly conceded that here in fact was a point of interest after all.

When the blaze on the tree was opened up fully to reveal the remainder of the markings, they were to be seen as plainly as the day they had been carved. They had been protected from the elements for so long that the wood had not been allowed to harden, as was the case with the 5. I knew that the now-exposed portion would rapidly degenerate until nothing remained.

In chopping open old survey blazes, as we had done so often in the past, we always cut a deep groove into the overgrowth, well clear of the estimated position of the concealed carvings. When the cut is deep enough, the slab will fall away easily after leverage with the axe, often in one piece. Then, as the new growth takes place, the slow creeping of the bark and wood will once again follow the contour of whatever lies beneath, faithfully reproducing every microscopic blemish as nature endeavours to heal the wound. Therefore, when our gaze finally dropped to the loose slab lying face upward at the foot of the tree, a further thrill was waiting for us. A perfect mirror image of a T over the figures 89 was in full view, standing out in bold relief from the curved interior of the hard bloodwood surface. Although from that moment onwards the years of the original carvings on the trunk were numbered — sad as it was to admit it — here was a permanent relic that would last forever.

I had gathered all our spare onion bags from the ration truck, and I spread them out by the tree, reverently placed the treasure on them, and wrapped it up carefully. I would carry it in a safe position in my Rover for months until finally returning to civilisation, where I could relate the story to unfortunate city dwellers who couldn't come out to see for themselves. William Tietkens celebrated his twenty-ninth birthday while on the Giles expedition in 1874, so this would make him forty-four years old when he had carved this tree.

After the road had been widened and graded, we moved our camp up to the point where the dozer had turned around, and each member of the camp was forced to stop and marvel at the discovery whether he wanted to or not. Most of the boys were impressed but again old Paul and Quinny stayed only long enough to satisfy me before continuing on to our new campsite.

Knowing that in less than a decade the markings except for the 5 would probably be lost and unreadable, I had long before decided to leave a signpost behind. This would be one of the usual aluminium plates nailed to a mulga post, with the exact information I had seen with my own eyes recorded on the tree punched on to its surface. With a cold chisel I fashioned the shape of the blaze around the letter and figures left by Tietkens, adding a note with the alphabet stamps to record all that was plainly visible when the new road progressed past this point. I finished with the date, 29 June 1960.

The camp on that freezing cold night was now 350 kilometres from where we had tapped our original Gunbarrel Highway, thirty kilometres east of Giles weather station.

Quinny and I now stopped at the tree to wait for the arrival of the bulldozer train, and while he sat in his cabin reading, I walked over to relive the story surrounding this area. So as not to waste time I opened my ever-ready pot of white paint and brightened up my signpost beside the road. As usual it had had around its base one of the old Land Rover's staked tyres, also painted white. In due course the rumblings drew nearer and the dust cloud became larger until it was time for us to move on.

We would soon be crossing over the Tropic of Capricorn, where yet another aluminium plate adorned a heavy mulga post complete with lead capping to inform travellers of the fact. Although this sign had no particular use, it was at least ornamental, and I had erected it to add to the many other points of interest this particular road had to offer. As we were building the road north of the Davenport Hills, we neared the latitude of 23°30′, and the thought had occurred to me that this was the value of the Tropic of Capricorn. I just could not pass over it without leaving a permanent monument at the point where the road turned tropical. An observation to the sun helped bring me near enough to carry out an accurate

TOP: Comfortably bedded down in the middle of our new road: no traffic to worry about yet. BOTTOM: The fantastic outcrop of red boulders near the Kintore Range. In the distance is Mount Leisler

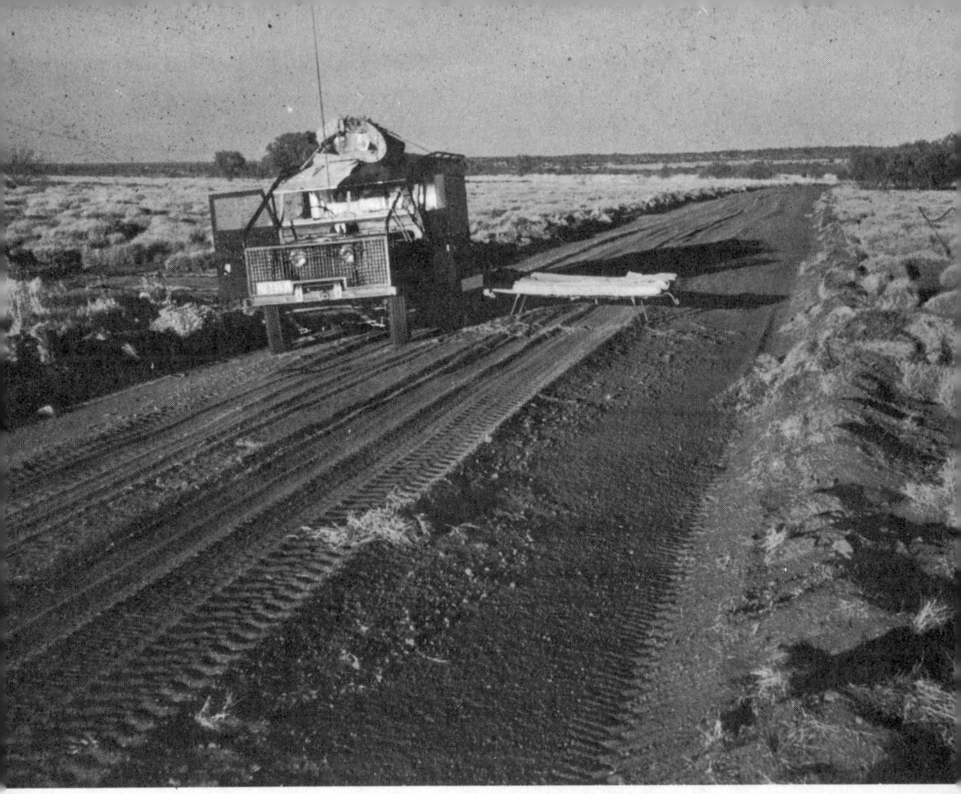

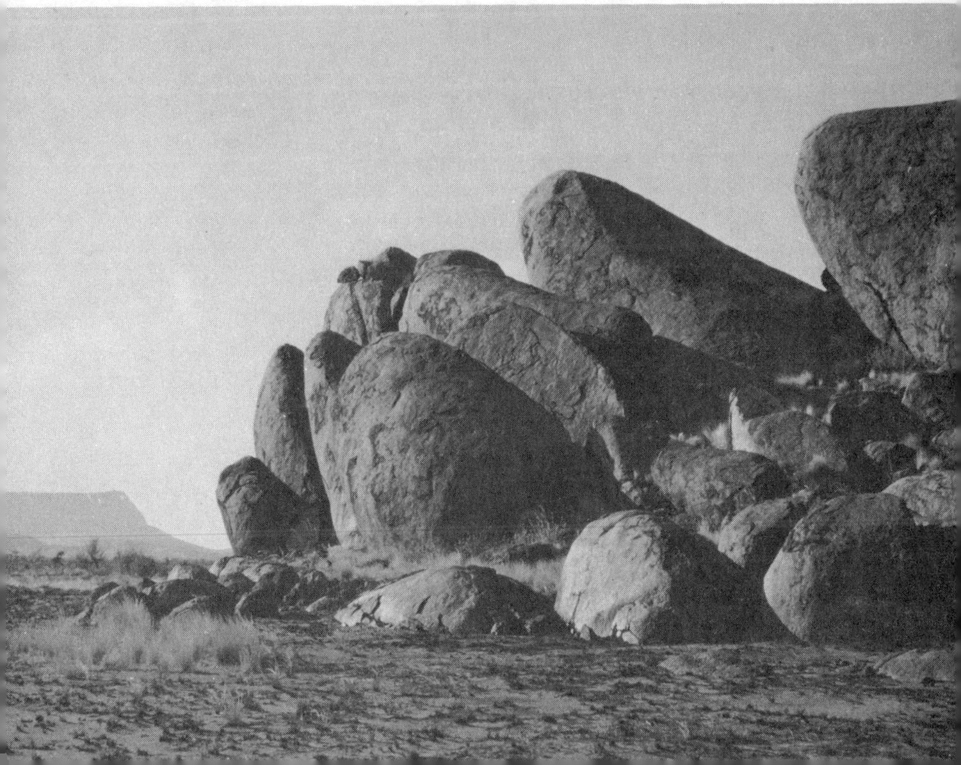

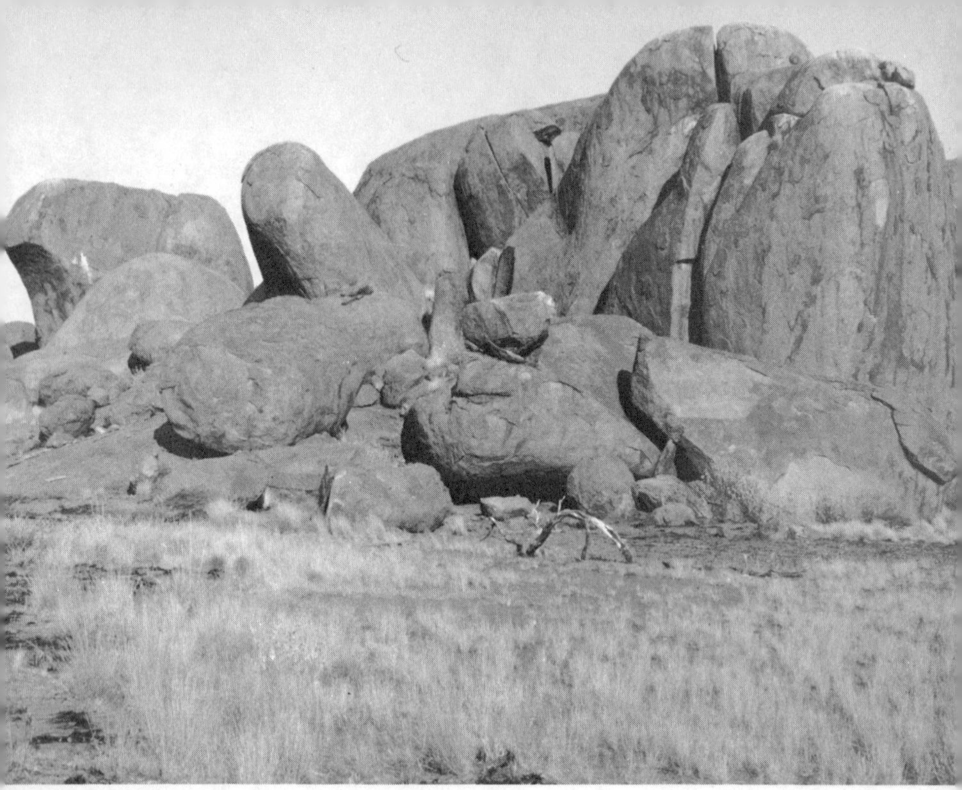

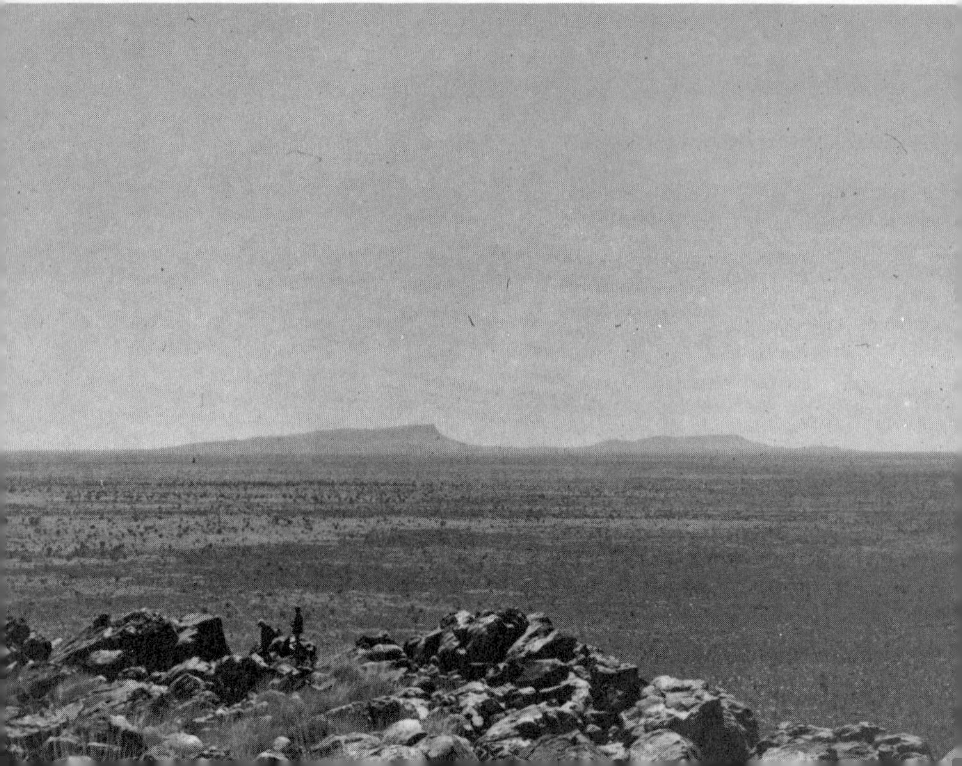

astrofix from which I could arrive at the spot to place the sign. Actually any latitude and longitude readings obtained on these roads were used by cartographers to produce new maps, and as all my figures were supplied to them, it wasn't entirely wasted.

I well remembered the cold, windy night when I stood at the theodolite to observe the stars at this point. Being late June it was cold enough already, without the exasperating stiff wind which, although not strong enough to shake the instrument, helped numb my fingers until they could barely turn the screws. When I had finished, late in the night, I packed up and returned to camp to sit huddled up in the back of the Rover, calculating the results. I meant to fix the signpost before moving from the region of the astronomical station, and so while Eric and Scotty went off the following morning to look for a large mulga post, I hammered out the aluminium plate. Another post on the other side of the road completed the effect I hoped to have on travellers, which was to make them think they were actually crossing over a line into a new climate. This done, they were both painted the inevitable white.

The point was a distance of 328 kilometres from the start of the road, and we had crossed over this invisible latitude line of Capricorn in longitude 129°20′38″, which information all appeared on the aluminium plate for the edification of passers by. Even Quinny's dog, Lassie, had her name recorded in the list of members of our camp at the time.

For tea that night, old Paul with a flourish had opened a tin of tropical fruit salad to celebrate the occasion. He insisted he felt decidedly warmer now that we were in the tropics.

Six months later our train groaned its way south between these posts and now, at this fiery time of the year, we were all pleased to leave the tropics behind. But as Ernest Giles would have said, "Herr Gabriel Daniel Fahrenheit still did not (con)descend to fall below a warm 108 degrees."

Six kilometres further south of the sign, and almost within sight of it, a huge rounded boulder was to be seen gleaming white through the ground heat shimmer, forming yet another monument at the roadside. Quinny and I, as usual in front of the convoy, stopped here to repaint it while the opportunity

Top: Another view of the remarkable boulder outcrop. Bottom: Mount Leisler as seen from the vantage point of the Davenport Hills

presented itself, even knowing that after the bulldozer train had passed by, the dust pall would again coat it in chocolate-coloured dust. This rock, weighing almost a tonne, was located exactly on the 322-kilometre point from the start of the road. This fact was made meaningless by the introduction of the metric system, but at the time of construction it was quite important. That monolith marked the even 200-mile point on the road since leaving the Gunbarrel Highway and served us as a landmark on many occasions, such as the time when the ration truck failed to return. A radio message would state whether the rock had been reached or not, and this helped us plan how far back we had to go for salvaging operations. Actually it had all started out as a joke, when we wondered how future travellers might puzzle over how such a large rock just happened to be so neatly and conveniently placed alongside the road at precisely the 200-mile point. A sort of little Stonehenge in the Australian desert. Nobody would ever know that we had placed it there ourselves.

As we had approached this even mileage, I had thought that we might erect at the spot a signpost with the information attached, but as we were very close to the tropical crossing, which would mean so much more, I decided against it. When the first cut of the bulldozer was made I did mark the spot with a small cairn, but no sooner had we passed the spot than the idea of the rock came to me. A kilometre north of the cairn, a series of submerged rocks needed to be negotiated right on our line. Not wanting to upset the long straight section of road aiming directly at the eastern highest bluff on Mount Leisler, we marked out its course over and beyond the outcrop. As the rocks could not be left where they were, the bulldozer returned to them and we began the operation of moving them out of the ground. With the slight movement they had made as the machine drove over them, this procedure appeared possible. Had they been joined to exposed bedrock, then a detour around them would have been necessary.

The great steel blade had gently ploughed into the ground just short of the first boulder and lifted half of it out at the first try. One after the other followed easily, although each must have weighed a tonne; and the next move was the job of filling in the resulting holes with blades full of desert. As this was going on I noticed that one of the largest rocks had on its otherwise rounded shape an unusual flat surface upon which a sign could be painted. This was it! Huge as it was, I was sure we could somehow get it to the 200-mile point. When Doug had repaired the road to a stage where Scotty could grade it, I had a conference with him. We decided that if we used the steel tow cable carried on the front of my Rover, we could actually tie the great rock on to the bulldozer blade, which, when raised with the powerful winch, should also be able to lift it clear of the ground. If we were successful, then the whole thing could be driven slowly back to what we hoped would be its final resting-place. After lowering it on the spot we could use the bulldozer to nudge it carefully into position, with the flat surface facing the road.

As always Doug was eager to carry out such jokes, in this case our "operation Stonehenge." After roping the rock into place we found that although the winch was able to raise it, the whole machine almost toppled forward again with the weight. It wasn't a recommended practice, we thought, after consideration of the front track roller and idler bearings, but out here on the spinifex plains nobody could possibly worry. After a very slow trip back, travelled in reverse, the load was eventually lowered, much to the relief of the bulldozer, and pushed into place. The rest of the camp had arrived during this exercise, and Paul came over to ask how long it would be before we had a spell from the bush. He was sure it was getting to us.

It took almost a tin of white paint to "ice" the rock, and afterwards it resembled a huge round cake. On a subsequent trip, after the paint had dried hard, I would sign-write in black a bold "200 Miles" on the flat surface which "happened" to be facing the road. I have since been asked by puzzled travellers just how we managed to make the road conform to those figures so accurately; so the whole operation had the desired effect.

We were now in the hemisphere south of the tropics, and the sun lit up this great white mass like a jewel on the spinifex prairie. As Quinny and I waited at the rock, I could be sure that this little joke of ours would soon be relived by Doug, driving that same bulldozer at the front of his dust-enveloped train.

From here the Davenport Hills could be clearly seen at a distance of only a further six kilometres. These, our next port of call, were the last of a succession of relatively small hills we had encountered before reaching Mount Leisler with the road.

It had been about here that Eric, who usually drove the workshop Rover, had begun to be affected by a soreness in his throat which slowly worsened by the month. At first only a slight swelling was noticeable, but, far from easing, this had enlarged to alarming proportions. He seemed sound enough otherwise and repeatedly declined the offer to return on one of Quinny's supply trips to have it seen to, as I guessed he suspected that this might put a stop to his activities with us. The prospect of leaving this work for good and being subjected to possible operations was not a very inviting one, and I knew at the time that I would strongly have resisted any such idea had it been myself.

While nothing really affected him or his activities we gave the swelling a chance to go away by itself, but the fact remained that something was very amiss. After a couple of months even Eric was thinking more about it, but being a former share-farmer he doggedly carried on until a visible loss of weight began to take place. Usually well built, solid, and the picture of health, he now contracted an associated stomach disorder. At first I could make no headway with him in making him see the sense of returning for expert help, but he now at last agreed that it might be advisable. Of course, I knew all the time that I could send him off even if he didn't agree. The only thing that forestalled this action was my understanding of how I would have felt in his place, and I wanted most of the decision to come from him.

It wasn't until four months after we had noticed the first symptoms that we were all convinced he should leave our camp on the very next ration trip, which I arranged to be made direct to Alice Springs. We had progressed with the road almost 300 kilometres west of Sandy Blight Junction, and as it was the last day of October, with the blazing heat already upon us, we were all aware that our year's work was drawing to a close anyway. Rolling up his swag for the last time in our camp, Eric weakly climbed up into the ration truck and it pulled out back along the new road to the east, bound for Alice Springs and the doctors.

After our long time together in this remote part of the country, it was depressing to lose one of our members, but we all knew as well as Eric that it was for his own good. When he eventually arrived in the Alice, Quinny had driven him straight to the hospital. A message sent over my transceiver was already waiting with the medical staff there. The resident doctor at Woomera, with whom I had already discussed this case, had on my request arranged a signal worded in medical terms to be transmitted to Alice Springs. I also sent a letter with Eric, outlining the circumstances which had brought about our decision for this step and describing what had transpired over the previous few months.

I heard later that the doctors at the hospital had taken one look at Eric and the large swelling on his throat and wanted to operate there and then. They spoke of thyroid glands and long hospitalisation, but Eric requested his return to Adelaide first so that he would be closer to home before all this took place. They reluctantly gave in and alerted the doctors down south that an urgent case was on the way and that no time was to be lost in effecting treatment on arrival.

The news had come back to me out at Jupiter Well, over our Woomera communication base station with whom I kept in much closer contact than ever before. As soon as the plane landed in Adelaide an ambulance had whisked Eric off to hospital, and I believe he was operated on that same day. I relayed the news to the camp after packing away the aerial. We all thought that we would never see him again in our little camp and I personally was very sorry about that, as he had always been a very willing and able worker, and easy to get along with. His background enabled him to turn his hand to almost everything, including assisting Rex with the big repair jobs on the machinery and driving the grader for a spell after Doug had damaged his hand. Apart from this, he always helped Paul in his daily chores of chopping firewood and packing.

Happily enough these gloomy thoughts proved to be short-lived, thanks to modern medicine, which could now handle his condition in a relatively simple manner. A book in old Paul's cook truck "library" described the trouble in detail, stating that in the not so distant past, operations to correct these swellings involved removing two glands. The unfortunate patient was often left a helpless moron as a result. Now, however, the same beneficial effect could be achieved by only partial removal, and the complete recovery of the patient was more or less a certainty.

So it was that as soon as the operation on Eric had been performed, his troubles were over, and he regained his former health and weight in an amazingly short time. In the years to follow, he returned to the bush with us and carried on as if nothing had happened.

Chapter Seven

THE DISCOVERY AT DAVENPORT

The first time I had ever visited this small but exciting cluster of hills named by William Tietkens after Sir Samuel Davenport, president of the South Australian Branch of the Royal Geographical Society, was during a forward reconnaissance five months before. Black clouds almost on the point of rain had turned the usual brassy skies into a blanket covering the desert to the horizon in every direction, and this, far from making the scene gloomy, added to the mysterious appearance of the outcrop. In the topmost crags, above the loose collection of huge rocks which had weathered free from their normal locations on the cliff faces, were dark cave openings which could be seen easily from the level of the plain. Some of the smooth monoliths had as yet retained their lofty positions, withstanding the millions of years of battering by the elements and stubbornly refusing to join the others at the base.

Our camp at the head of the road to date lay sixteen kilometres back across a flat, scrubby plain which was bounded on its northern and southern edges by long narrow sand-ridges. The belt on the southern edge extended for over 600 kilometres from east to west, and we had been battling with it for weeks. At last, after negotiating only three more sand-ridges, the long grind (although we didn't know it at the time) was temporarily over. We had become entrenched between two high parallel ridges 100 metres apart, running at right angles to the direction in which I wanted to take the road. There was no hope of crossing over them, so the road necessarily had to follow their course until we could find an outlet to the north. In this sort of situation, which occurred quite frequently, I likened our bulldozer to a gramophone needle, caught up in a groove from which there was no escape.

Climbing to the top of the ridge on our northern side, I could plainly see the Davenport Hills, so near and yet so far. Beyond them reared the unmistakable bulk of Mount Leisler. If only I could break out of this slot between the sandhills, there didn't seem to be much more to prevent us reaching them with the road, but until I was actually on my way there I couldn't be sure. With this in mind I left the camp to recover from the previous beating it had taken, both mechanically and physically, and set off on another forward survey, as usual alone.

Although I could often reach my destination in the Rover, driving across sandhills and rocky outcrops, between powdery white gypsum patches and over low-lying potential swamps, this didn't help. My aim was to find a feasible route for the future road, leaving wheeltracks over which I could drive again to direct the dozer. So, setting off from the end of the present road, I allowed the high sand-ridges to govern my course, at the same time keeping an eye to the north for a possible escape from the valley.

As is often the case, the mulgas and scrub between the sandhills were reasonably thick because the infrequent rain naturally flowed to the lowest point. The seeds were washed down and subsequently watered by the several showers per year which visited this area so far removed from the sea. It was hard going plunging through all this growth, so I often skirted it along the lower reaches of the sand on the northern side of the valley.

Surprisingly, after driving for about half a kilometre, I was elated to see that the seemingly never-ending sand-ridge abruptly petered out to nothing. The level of the valley continued on unobstructed around the terminus, leaving the way clear towards the Davenport Hills. Although still scrubby, the terrain was at least flat. We were out of that dreaded belt of sandhills at last. From years of past experiences, I didn't allow myself to become too overjoyed, for I knew they would rear up in due course to bar our way once more. Still, these temporary lulls were always very welcome.

I headed off in a direct line towards the highest of the Davenport cluster of outcrops, and after rounding the point of the sandhill, I endeavoured to keep the wheeltracks as straight as possible. I did this in the hope that the final location of the road would be over these tracks, and that nothing more would loom up to prevent it. Everything went well for over nine kilometres, with the rocky hill in clear sight all the way, until inevitably another long east-west sand-ridge began to appear. Nevertheless, that previous straight stretch would surely govern the new road and I would come to grips with this new obstacle in one way or another.

Then to the west I noticed another sandhill. It was removed from the first by the usual valley width, but this one had already come to its end, leaving the flat country unmarred. I soon deduced that by some rare and most fortunate stroke of luck, the straight line course across the scrubby plain had in effect stretched exactly from the first terminus to the second. No correction was needed at all, and I sedately curved my line to pass into the valley between the two, this time heading west. This kind of fluke happened so infrequently that I was still feeling pleased with it as I trundled over the spinifex clumps which had replaced the usual thick bush in this particular groove.

By now the rocky hills appeared to be only just over the other side of this last ridge, so all that was left was to clear it and the new road leading right up to them would be as good as made. For the first time in history the Davenport Hills, last visited by William Henry Tietkens in 1889, would be accessible by a made road from the south.

In a matter of less than a kilometre the main ridge blocking the way on the north side fell away to plain level, leaving a clear path to the hills. Actually, although I didn't know it at the time, the barrier was formed by two parallel sand-ridges of uniform height, and on rounding the point of the first, I saw the other which had also come to its end in the same area. At this stage I had been considering working on the sand with the bulldozer, reducing its height for the road instead of skirting it, but the sight of the second termination dispelled all thoughts of that. As I closed the short distance to the foothills, which had never before had a motor vehicle driven over them, I decided that this was my lucky day.

As always, whenever in any proximity to such points of elevation, the first impulse is to climb to the summit to obtain a bird's-eye view of the surrounding topography. In my case it helped to form a better idea of where our new road could be located, so, leaving the Rover on the lower slopes among the spinifex, I began the relatively small climb.

Because it was a dull, grey day, with heavy cloud cover forming a complete blanket over the scene, I left my camera in the vehicle for once and set off with binoculars only towards the first of the visible caves. I was not yet to know how much I would regret not bringing the camera with me.

The only sound to be heard was the occasional scraping of my iron-heeled boots on the rocks, and in a comparatively short time I had gained a ledge just short of the top. As I was quite alone, no conversation resulted when I sat on a boulder to catch my breath and scanned in complete quiet what I could see of the horizon. Suddenly, where the skyline was cut by large rocks even higher up than my ledge, a movement caught the corner of my eye.

At first I couldn't believe what I saw. At close range the furry shape of a small animal was sitting at the entrance of a cave, silently contemplating me without the least sign of fright. It had black and white markings down to its nose, which it stroked with its forepaws, and in the sitting position its white front showed towards me. I estimated its overall height to be about forty centimetres. A black untapered tail equal in length to the height and about five centimetres in diameter emerged from the base of the blue-grey back to hang down over the ledge of rock on which the animal was sitting, and two rounded bear-like ears nestled close to the small head.

Carefully taking out the binoculars, I was able to focus on the little creature at a distance of only ten metres. Through the glass it seemed as if I could almost touch it. Never before, after so many years of exploration and surveying in extremely remote areas of Central Australia, and climbing hundreds of such vantage points, had I ever encountered an animal such as this.

My mind, searching for an explanation, almost immediately turned to a phrase I had read in Ernest Giles' journal of 1874. On his expedition into the regions only 150 kilometres south of where I was sitting at that moment, he had attempted to cross Australia to the west by horseback. Now, with our knowledge of what lay ahead of him, we know this would have been impossible, and he himself proved it; but at that time such knowledge was yet to be acquired. After he had lost Gibson to the desert and had virtually crawled the 200 kilometres back to where Tietkens was waiting at the Rawlinson Ranges, he was on the point of exhaustion, not only from a choking thirst but also from extreme hunger. For over a week he had not eaten anything but strips of smoked horsemeat cut from one of his animals, Terrible Billy.

Tietkens and Andrews, the other members of his expedition,

were camped at a rock-hole he had named Fort McKellar on the northern side of the Rawlinson Ranges, waiting while he and Gibson attempted the crossing. Thirty kilometres to the west of the Fort was another rock-hole they had named the Circus Waters—the last supply known to them before plunging into the howling wilderness of what is now the Gibson Desert. I myself had re-found the Circus Waters several years before and obtained a small quantity of water from the same rock-hole, this being its first visitation by whites since that fateful trip of Giles and Gibson.

When Giles reached the Circus on his knees he lay down and drank for hours. Even in his delirium he realised he still had to drag himself a further thirty kilometres on to the Fort. His plight can be understood, for he discovered that Gibson had not been there, and before being able to return and search for him he must first urgently get back to Tietkens and the other horses. The driving force of this need provoked his next actions, and the relevant words from his journal were now repeating themselves in my mind on the Davenport Hills.

"I was very footsore, and could only go at a snail's pace. Just as I got clear of the bank of the creek, I heard a faint squeak, and looking about I saw, and immediately caught, a small dying wallaby, whose marsupial mother had evidently thrown it from her pouch. It only weighed about two ounces and was scarcely furnished yet with fur. The instant I saw it, like an eagle I pounced upon it and ate it, living, raw, dying—fur, skin, bones, skull, and all. The delicious taste of that creature I shall never forget. I only wished I had its mother and father to serve in the same way."

The little creature at present in the field of view of my binoculars was surely of the same family.

With my camera down at the bottom of the hill in the Rover I could only construct a mind's-eye picture of the details, but I determined to draw a sketch of what I remembered as soon as I regained the vehicle. It would be the best I could do in reporting the find to interested bodies in Adelaide and having it identified. After a long examination I decided to attempt an even closer look, and, returning the glass to its container, I slowly moved forward. Never having seen a human being before, my little furry friend still showed no signs of distress as it contemplated my movements, but it obviously couldn't understand the slight noise of my hobnailed boots on the rocks.

Continuing to stroke its nose with the black paws, it allowed me to halve the distance between us before it bent over, manoeuvred itself around, and in leisurely fashion walked on all fours back through the entrance of its shiny-walled cave. My next step was to scan the vicinity for others, and I was soon rewarded by seeing no fewer than six of them spread over an area of about a hundred metres. One, silhouetted against the grey sky, was perched on a higher boulder observing the proceedings, and others were ambling along rock-ledges, emerging from and re-entering their caves.

At last I ventured right up to the nearest cave into which the first one had disappeared so that I could examine at close range the mirror finish of the walls. The movements of these animals, brushing against the rocks for countless centuries, had obviously produced this silky smooth and gleaming effect, and I was sure that if it had been a usual sunny day the surfaces would have reflected light brilliantly. Higher up and out of reach the rock-faces resumed their usual rough, dull grey appearance, and the level floors were in places worn into hollows from endless comings and goings of little paws.

One of the caves was slightly larger than others, and I was able to crawl in to a distance of several metres until even the sparse light from the cloudy sky became blotted out altogether. I resolved to return to it as soon as possible, armed with a torch, to reveal what had been hidden from human view since the world began.

At last I continued my climb to the topmost rocks, to be presented with a full circle of unbroken skyline. I was very pleased with the inspection of the intervening country between there and Mount Leisler; it appeared to be entirely free from sand-ridges for the entire distance. Even through the field glasses and with the absence on this particular day of the perpetual heat shimmer, nothing but a spinifex plain dotted with occasional shrubs could be seen. My new road would clear the terminus of a short sandy spur trailing off the rocks at the base of the hill towards the east, then run in a dead straight length of twenty kilometres aimed at the highest point of Mount Leisler.

The spectacle of the blue bulk of Mount Leisler at this distance made me feel indescribably elated with the knowledge that the location of the next thirty-odd kilometres of road from my present camp was now a foregone conclusion. Excited as I was at my latest discovery, and having come so close to the goal I had set myself with that first sighting of the mountain so long before, I felt on top of the world. The phrase which describes something about somebody being "master of all he surveys" loomed up now in my thoughts, and I could well understand the emotions that had prompted such a remark.

The unbroken sea of sand-ridges through which we had weaved and bulldozed was in stark evidence towards the southern skyline, but that particular battle had already been won. I knew we had thousands of kilometres ahead yet to conquer and, judging by the thousands of kilometres already covered by our roads, I also knew we were by no means out of it. Breaking the whole into a series of minor goals, as I have so often said, is the way in which to tackle any project of gigantic proportions, and those thoughts in no way detracted from my present feeling of exhilaration.

Turning my attention once more to the sketch I must compile of my new-found friends before the memory picture faded, I reluctantly began my descent from this most inspiring place to carry on with it. The general atmosphere of the area was also becoming darker, if possible, and this indicated that the sun above the heavy cloud blanket was beginning to set. I would have to start the return trip to camp before complete nightfall. I still had my wheeltracks to follow back as these, in this case, need not be altered in any way.

Once in the cabin of the Rover I took up a soft pencil from the little tray I had bolted to the door for such things and opened a field book at the last page. From memory I was able to produce a reasonable likeness to what I had observed through the glasses, and when I had finished, I carefully packed it away until I could show the experts.

When I eventually did have the chance much later to

discuss it "down south," my drawing was immediately recognised as representing *Petrogale lateralis* or Black-flanked Rock Wallaby. The species used to exist in large numbers in the arid sections of Central Australia, inhabiting mainly the rocky crags and elevated sections of mountains or high outcrops. In fact, Ernest Giles also remarked that on an earlier occasion Andrews and Gibson had succeeded in shooting three wallabies, so they must have been more prevalent even then. In recent times fewer and fewer had been reported, for their population had been greatly reduced by foxes, dingoes, and possibly Wedge-tailed Eagles. This depletion had been so pronounced that during the quarter-century I have worked in this country, I had never encountered one before, nor have I since. My find and subsequent reporting made a two-column article in one of the Adelaide newspapers, and a reply by a zoological expert was published the following day.

Since I made the discovery, fleeting glimpses of these rock wallabies have been made by others on visits to the summit of the Davenport Hills, but no further sighting has been made at so close a range or for so long a period. Anyone who has accompanied me on subsequent climbs has been advised to maintain complete silence so that the conditions of that first lone and quiet meeting may be recreated. Knowing that the little animals are inhabitants of the upper crags and pinnacles, we have always treated that place as a sanctuary. Normally they shelter in their polished caves from the heat of the day, and emerge to feed at night. If my first trip had not been made while the sky was heavily clouded, giving a semblance of darkness, I would never have seen them at all, any more than if I had had noisy company. I speculated that, given an exactly similar set of circumstances, I might have had other meetings with them over the years. It is true that I was mostly alone on such climbs, but as a rule the scorching sun which seared the country for over ninety per cent of the year would have kept the little creatures well hidden in the dark. Apart from this, not many hilltops were embellished with those attractive caves which seemed perfect for their habits.

With our bulldozer train at the end of that long straight section

Top: The striking conformation of Mount Leisler rears up in marked contrast to the surrounding flat spinifex plain. Bottom: Tietkens' tree. Discovering this historic bloodwood was one of the highlights of building the Sandy Blight Junction Road

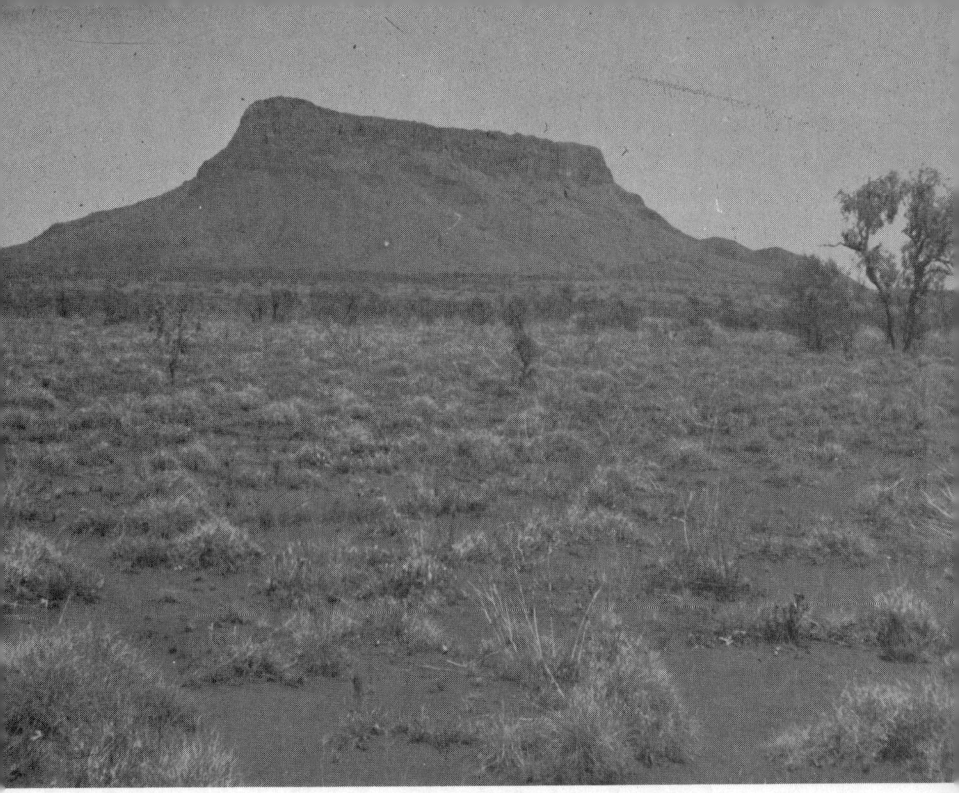

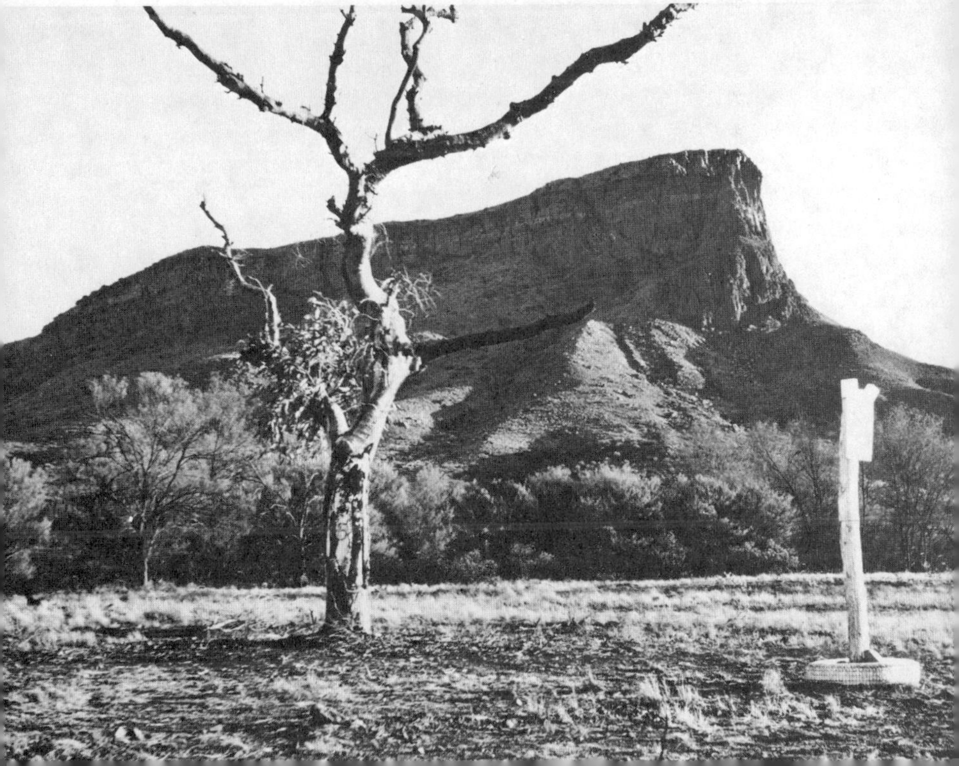

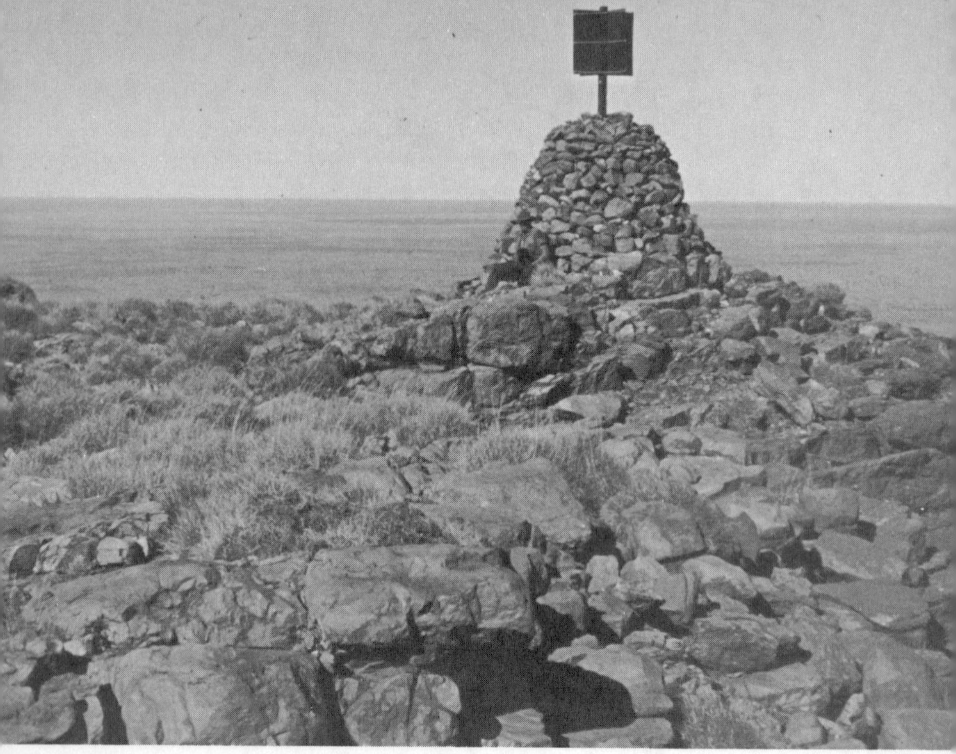

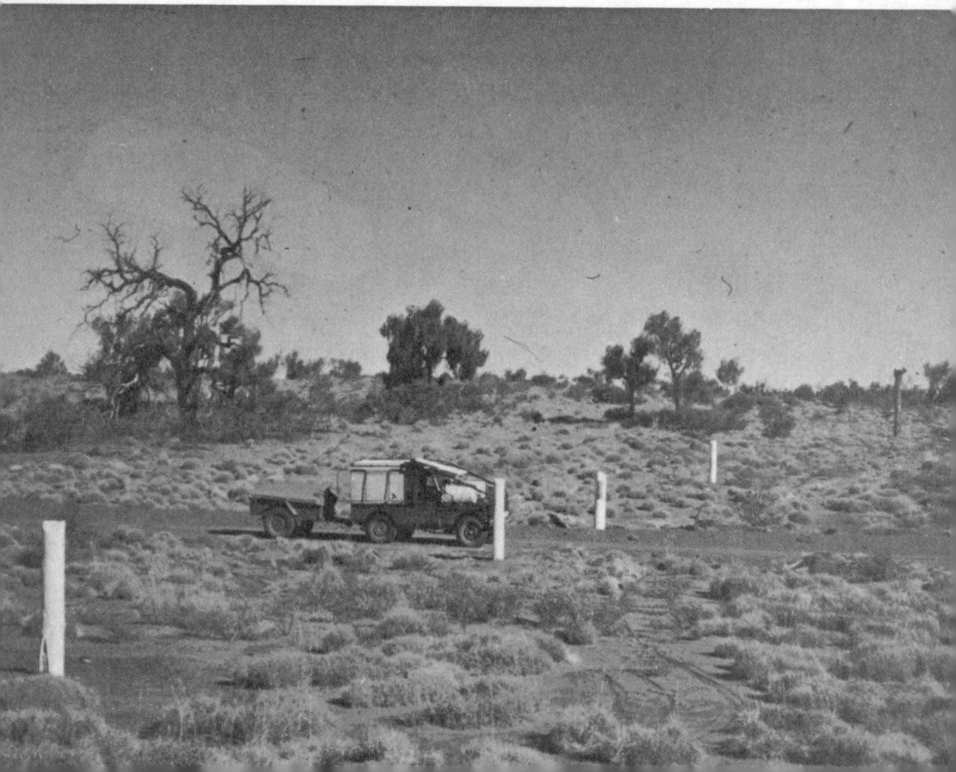

of road, we were now able to round the point of the sandy spur leading away from the eastern foothills. This meant that within a few hundred metres we would be noisily clattering past the base of these Davenport Hills, with no hope of seeing our friends. That they were still there was a certainty, but as this ear-splitting cavalcade thundered slowly past the lower reaches of the slopes where I had left my Rover on that first eventful visit, they would all be well concealed in their caves. In addition to the noise, the blistering sun would have kept them "indoors" as it had for centuries before our road invaded their privacy, ensured up till now by their extreme isolation.

Even though Quinny and I had reached the area in comparative quiet, well in advance of the rest of the convoy, we didn't detect a movement as we lit a fire in preparation for our overnight halt.

TOP: The trig survey beacon later erected on the summit of Mount Leisler, with Quinny in front. BOTTOM: We used white posts to mark out the border crossing between the Northern Territory and Western Australia

Chapter Eight

AN OPEN-AIR WORKSHOP

This particular camp had been the scene of an incredible amount of bush mechanical work when it had also marked the head of the road during construction. It had been almost winter at the time, and the usual spate of broken springs was upon us. The unfortunate spring assemblies, already of necessity grossly overloaded, would become freezing cold in the winter months, and this did not allow the flexibility required to cope with the rough going. They snapped like carrots as the vehicles were driven over dry creek-beds, clumps of spinifex, and exposed ant-beds, and with the added exertion of the heavy loads taking most of the shape out of them, not much was left in reserve for when they were needed most. The result was evident almost every time an under-vehicle inspection was made. As this scrutiny took place daily, broken shackle pins and mainsprings, curled leaves, and other problems were discovered with great regularity, and the eventual repair or replacement work would be saved up for camps such as this. In the meantime, chopped-off mulga branches would be wedged in between the axles and chassis to keep the damaged vehicle on an even keel, and these would be wired into place with the usual bush standby of fencing wire.

Then for a long time there had been a glaring need for a fitting to help rip out ant-beds from the surface of our new roads. White ants often construct huge above-ground castles in which to live, but in this section of the country the only evidences of their presence were bare patches on the surface. These were far too plentiful to avoid, and as the bulldozer with its bladeful of dirt and shrubs skimmed the ground, the submerged ant-beds were left on an equal level with the intervening red laterite. The beds were iron-hard, and if the leading edge of the blade happened to be only slightly under the surface, it would be forced up and over the structure, not even marking it. In fact, had there been enough ant-beds to make a continuous line, a perfectly hard, instant road would have been left, impervious to water and strong enough to support the heaviest vehicles. As it was, they were awkwardly placed several metres apart, sometimes covering thousands of square kilometres at a time, with areas of soft soil in between. Some sections of the plains and valleys between sand-ridges were worse than others, and the finished road over these areas would soon prove to be the roughest, at its worst a real spring-breaker.

After the grader had finished, the smooth ribbon of road strung out behind appeared to be ideal, a terrific contrast to the scrub, dead branches, and uneven hummocks lining it on either side. Of course some of this beautifully flat surface was made up of ant-bed tops, and the remaining sections between were soft, powdery red dirt. The first vehicle over it would start the rot, and this would be added to by each successive trip as the wheels would sink into the soft areas and ride up over the beds. In an alarmingly short time the "perfect" road would be reduced to hills and hollows with about ten centimetres' difference in height, and then the hills would take on vertical walls as the wheels, big or small, would accentuate this unevenness. Finally it was almost better to drive alongside the road than on it. The built-in scarifier on the grader could cope with the ant-beds, laboriously ripping at their surfaces with its comb-like tines, but as it attacked the ones on the shoulders the vehicle's outside rubber tyres would be riding over the windrow among the dry stakes, and many flat tyres resulted.

The answer seemed to be either to extend this scarifier to the full width of the machine, or to incorporate some sort of teeth on the bulldozer. Widening the scarifier proved fruitless, as the weight and traction weren't there to push the extra load. After we had tried it we conceded that the Caterpillar people knew what they were doing after all. Once the rock-like ant-beds were teased up near their tops, the debris could be spread along the road to amalgamate with the softer soil thus forming quite a uniform texture. This material was capable of even compaction, and as we had found over the years by persevering on the worst sections with the grader, a permanent good road was left. The disadvantages were, firstly, the time it took to concentrate on comparatively small lengths of road, and secondly, the great wear and tear on the big grader tyres.

As this trouble had plagued us for many years over thousands of kilometres of new road, resulting in not only dozens of broken springs on our supply truck but also in its load being shaken to pieces, we were constantly inventing new ways to combat it. Sometimes, in order to make the regular supply trips back along just-made roads less of a torturous ordeal, we saved up the task of tining and regrading the more wretched sections for later, whenever the opportunity arose. This might happen when I would be forced to carry out a long preliminary solo expedition in order to discover what lay in store for us in this unmapped country. I could then plan the course of the road hundreds of kilometres in advance so that future travellers would not have to waste time and fuel.

At such times, if any large-scale work was needed on any of the vital parts of our machinery, an opportunity might present itself to have the grader and one vehicle with supplies dispatched to some trouble zone. Scotty on the grader could then make our future supply trips a great deal easier with what time he had available before we were again ready to resume construction. Sad to say, these opportunities didn't present themselves often enough, or did so in places so far removed from the ant-bed infested sections as to make any sort of return impracticable. If only we could handle this battle with the ants on our first bash through using the huge bulldozer with its weight and steel traction plates, we could leave behind a much more desirable access. It was at this camp that we finally decided to attempt a modification.

Among our gear was an excellent arc-welding machine driven by a petrol engine, together with oxyacetylene torches, bottles of gas, and spare teeth for the grader scarifier.

The pleasant campsite among the desert oaks and under the shadow of the Davenport Hills thus became a hive of heavy bush industry as welding sparks flew and great chunks of steel were cut and fashioned into an ungainly-looking fitting. This was supposed to be bolted with high-tensile steel bolts from our supply carried for attaching new blades to the grader's mouldboard, leaving the tips of the reinforced tines protruding under the lower edge of the gleaming dozer blade. When completed it all looked like something out of the arc, but it seemed very functional, so we were eager to test it out on the previous straight stretch of road across the scrubby plain.

It worked all right. It not only teased the top few centimetres of the submerged ant-beds but it hooked on to them and lifted them bodily out of the ground with the action of a dental root elevator on a molar. With the sudden removal from the ground of a great spherical core measuring up to two metres in diameter, the logical result of a two-metre-deep hole was achieved. This in no way added to the silky-smooth result for which we had strived, and as they came at the rate of dozens per every hundred metres, a considerable amount of work was needed quickly to repair the damage.

Still, some of the long nightmare sections of bone-shattering ant-bed patches could do with such drastic treatment, as anything we did at this stage could only enhance the road surface. Nothing could make it worse. Because of these thoughts we weren't entirely disenchanted with our inventive efforts. Nevertheless, we promptly unbolted the device for use later on, in case it "might come in handy"—a descriptive phrase which was gradually being subconsciously applied to a large proportion of our equipment. The saying had been first coined by old Paul. Often, as we poked about in some pile of worthless junk near a station homestead on our way to the desert, Paul would amble over and advise us to throw on one or two items, adding that "they might come in handy." Periodically, after retaining our priceless hoard of "handy" items for months or even years, and guarding it with our very lives, we would reverently deposit large proportions of it in lonely graves in the desert, dug by the very bulldozer for which it was intended. Occasionally some diminutive item would come into its own and be the saviour of some big operation, but we often wondered if the constant loading and unloading of the rest of it over such long periods was really worth it.

Also at this Davenport overnight camp we had included such other work as the job of replacing a slipping clutch-plate in my Rover—a relatively large operation in such an environment. It had been the cause of lost power for many weeks since I first detected it, but the great bulk of the sand-ridge belt through which we had been pushing had certainly not put us in the frame of mind to tackle so large a task. But, as is always the case in almost all fields of work, pleasant or unpleasant surroundings and atmosphere govern the efficiency with which the jobs are handled.

With the Rover neatly parked out of the winter-cold shade of the desert oaks, Rex, with me giving what little help I could, dismantled the middle section of the little vehicle to reveal the offending clutch-plate. The job could only be started after hours of unpacking my assortment of apparatus necessary both for deciding the course of the road on the ground and plotting the finished location on paper. Theodolite and tripod, boxes of nautical almanacs and calculation books, padded containers for stop watches, radio transceiver for the reception of time signals as well as communications, rifle, revolver, camera, geological hammer, and tools were neatly piled on a large canvas camp-sheet alongside the Rover. Swag-roll of blankets, pieces of "handy" rope, lengths of fencing wire, axe, tuckerbox with billies and frying pan, torch, and a ten-centimetre roll of white crêpe paper were soon added to the swelling heap on the canvas. Of course the strong tin of spare engine oil in its rag wrappings and the dust-proof box of dental instruments, together with the engine spares, had to go as well; and then at long last Rex could begin his part.

As I gazed at the mountain of gear arrayed on the camp-sheet I wondered how on earth it had all fitted into my comparatively small vehicle and still left room for me to climb in. It was, as Paul said, amazing. Nevertheless, all these things were in constant use and made it possible for me to carry out my succession of lonely and lengthy expeditions into the unknown. Most of the items definitely came under the heading of "would" come in handy, instead of "might."

The floorboards and seat fittings were the first to go, followed by disconnected gear and brake levers, placed in an orderly fashion on the doors, which also needed removing to admit both of us freely. The tail driving-shaft came next, and all the bolts and studs from the flywheel housing were carefully preserved in an opened jam-tin, hopefully for replacing when and if the fault was located and remedied. Eventually the clutch-plate assembly was revealed, and I couldn't help realising how often my life had depended on that little disc. Everyone relies daily on such apparently insignificant things for their very existence.

When Rex finally held the plate in his hands, we both made our inspection and followed it with the same remark: there was nothing obviously amiss with it at all. It certainly wasn't worn or oily. In his usual calm, unruffled way Rex, covered in oil and grease, ventured a polite opinion that perhaps the lack of pulling power in my vehicle might be due to something else. With the strong little engine and its extra low gearing, I did find more and more that although the speed of the motor remained unaltered, the wheels were not fighting as they used to. In fact I often had quite a struggle to extricate the vehicle from dry creek-beds and gain enough momentum to accomplish the hundreds of sand-ridge crossings.

In any case, now the transmission was all in pieces, we decided that a careful use of the oxy torch might remove any stray lubricant which could have penetrated the plates, and so Rex patiently warmed and rubbed every square centimetre of the surfaces. After some adjusting of springs and levers, we began the task of replacing everything in its original position, and finally, as the last bolt was screwed into the floorboards, I started on the really hard part of the operation. Dovetailing all this mountain of equipment back into place was an art perfected only after years of practice.

Then came the mending of all the broken spring assemblies on a large percentage of the other vehicles. Quinny's supply truck had been subjected to almost as much suffering as my Rover on the thousands of kilometres of travelling for rations, water, diesel, and petrol to keep us going. While the big-scale job of removing his great heavy springs was progressing, I finished repacking my Rover and took it for a sandhill test.

In normal garages, repaired cars are taken for road tests, as there is no shortage of roads; and here we certainly had no shortage of sandhills. Whatever had been done in the processing of my clutch system restoration, or a combination of everything, had happily done the trick. With slackened tyres I literally flew over some of the smaller test mounds. I was once more ready to tackle whatever terrain loomed up ahead so that I could continue to guide the bulldozer, my one aim in life at that time. Everything we did went towards the single goal of keeping the dozer on the move as it had been for the previous four years, and, as it happened, for the following four years as well.

Paul's ration and cook wagon never seemed to need any attention at all. His two big journeys for the year took place at the beginning, on our way out to the desert to resume our work, and at the end, on our way back. For the remaining ten months his truck crawled along at an average of six or seven kilometres a day, keeping abreast of the head of the road as we forged slowly ahead. His daily move started long after Doug and I had departed from the camp for the bulldozer waiting at the extreme end of the cleared ribbon of ground. After breakfast we would head off, having loaded on the swags when we got around to "balancing on one end," as a former member of the party always described the operation of standing up. For some time before this Paul would have had the fire going and what breakfast was available cooking. The outfit was such a mobile one that there wasn't very much to pack up, and during the morning the retinue would slowly creep along the huge distance of half a dozen kilometres to the current terminus. This length was made up of the three or so kilometres we had made the previous afternoon plus the three or so we would have made next morning.

When the time reached an hour before noon, I would select a spot for an overnight camp with an eye to large amounts of dry wood close by in winter and any possible shade in summer. Doug would then clear an area alongside the roadhead with a few passes of his huge blade in as many minutes, and we would press on with our job. Paul would simply keep coming, and when he found this clearing, half the size of a tennis court, he would drive into it and in no time the billy for the dinner camp would be on the fire. While Doug and I had something to eat, Rex could carry out his daily service of the bulldozer (which was by this method sometimes right at the camp), leaving it greased and refuelled for us to resume the attack into the scrub ahead. The afternoon's work left the heavy machine at the end of the road, and Doug would return to camp the relatively short distance in my Rover.

The object of this constant moving was to obviate the necessity for churning over long distances to any sort of a standing camp. A newly-made bush road is soft, being yet uncompacted by traffic, and it takes years before the surface can be driven over with any ease. This "heavy" surface, together with the dead weight of our overloaded work-horses, reduces the kilometres capable of being travelled per litre of petrol to only about one. With nearest supply of petrol and everything else being often thousands of kilometres distant, this wastage would have added up to gigantic proportions over even a short period, and we simply couldn't afford to use up fuel at such a rate.

If a camp had remained in the same place for even a week, then by the end of that time the head of the road would have been forty kilometres away, necessitating an eighty-kilometre return trip to and from work. At the rate of one kilometre to one litre of petrol, as was the case especially with the fitter's truck, eighty litres would be needed just to service the machine. Then again, in open spinifex country where only one pass was needed with the bulldozer, leaving to the grader the job of widening the road to double the length of the blade, distances of sometimes fifteen kilometres could be accomplished every day. The travelling then entailed would be out of all proportion to the meagre supply of fuel we were able to carry, and it would make our task virtually impossible.

On the other hand, it occasionally happened that we would have to cross some particularly persistent sand-ridge which just would not come to an end, followed sometimes by others at close proximity. A slight reduction in the height, forming a natural saddle, might then govern where we would make the crossing, even though in itself it was still too high to negotiate. This would mean days of laborious work with the bulldozer to reduce its height to such an extent that we could grade a finished access over it. To prevent the resulting walls of sand at each side from drifting back on to the wheeltracks to meet each other in the middle, these crossings had to be of necessity anything up to half a dozen blades wide. As the first could take anything up to several days to complete, and subsequent ones just as long, the ritual of packing up would be unnecessary at such times, and the camp would remain stationary for as long as it took to carve out the way.

On my solitary advance reconnaissances, I would design the ultimate location of the road to be clear of such obstacles if at all possible and would sometimes labour the little Rover for fifty kilometres searching for an outlet. At times, after following a ridge in its adjoining valley for such a distance as to make a crossing (with an eye to my distant destination) completely impracticable, I would try the other way. Then, after tracing it back for a similar distance, I often found to my sorrow that the high sand barrier merely converged on another of equal height, blocking the way completely. This invariably meant attacking the first at its lowest point, and thus a week's spell from the daily packing operation.

It was on a survey such as the foregoing, several years before, that I rued the fact that I couldn't keep the road as straight as I would have liked. The picture of the corkscrew shape into which the future road would be forced by the topography it had to follow kept appearing in my mind's eye. The word "straight" repeated itself in my mind as I battled along over mammoth hummocks of spinifex, until a name suddenly materialised as clear as crystal in my brain. What could be straighter than a gunbarrel, and what better name could I give the little team I had dedicated to making straight roads where feasible than the "Gunbarrel Road Construction Party"? I remembered at the time returning to the camp and informing them of their new title. This name, of course, remained with us forever, even to the official label given to our first road across Central Australia being the Gunbarrel Highway.

Rex, between servicing everything else and helping Paul with the continual moving, always kept his own fitter's truck at peak performance, so he could allot all his time to the others. By the time he had battled with Quinny's troubles and mine, not to mention the grader and bulldozer ant-bed modifications, he could at last turn his attention elsewhere. Here at Davenport the broken speedometer cable on the workshop Rover needed to be taken care of, together with its cracked differential housing, and of course the leaky valve on the refrigerator compressor, installed on its little trailer behind the ration truck, must be remedied.

Now, as our clumsy, ground-shaking train ground to a noisy stop, I became mindful of the colossal amount of work we had done on it all at this very camp.

The fridge engine, although much the worse for wear after the fire, was at least still with us and working; Quinny's truck was still battling on, in spite of a few wounds which needed licking; and the valiant old bulldozer was quite unruffled by all the goings-on in the desert. My Rover with its non-slipping clutch-plate was as reliable as ever, as was the workshop Rover, complete with its welded differential housing and replaced speedo cable. The grader, the basic cause of this titanic towing operation, was at least still right here with us.

The sand around this overnight camp still showed signs of oil patches, a reminder of one of the most comprehensive programmes of bush mechanics ever carried out in the one place by our little team.

Ironically, the only vehicle which hadn't needed attention at that time, and the one least called upon to face up to the demanding conditions of work in our outback venue, was also the only one no longer with us. It was quietly reclining among its ashes at its last resting place, 240 kilometres back along our new road.

Chapter Nine

BUSH BEDROOMS

Over the years each one of us had developed his own scheme of sleeping arrangements, and even now, during this shattering deviation from our usual routine, we all put in the night as before, as if nothing had happened. Even old Paul, suddenly minus his faithful ration truck, was still to be seen lying out in the open on his iron, hospital-type army bed with its wire mesh top.

When this project had first begun, we had all camped on the ground among the saltbush prickles, thorny spinifex, and powdery bulldust. As our campsite was changed almost every night, we gave less and less attention to preparing a sleeping area, and only the sharpest rocks in stony places were thrown aside. In khaki-burr patches, some of the spiky little balls of needles would inevitably become embedded in the blankets, leaving behind hair-like needles even when the main bulk was removed. After rainy nights, even though the swags were rolled in protective canvas camp-sheets, it would be days before we could dry out the covers, holding them up to large fires before the following attempt at sleeping.

We made a big concession when, several years earlier, pillows were brought in lieu of heaped-up clothes and towels; but during the winter months the dew would gradually penetrate them, and after a shower of rain they became quite unusable. The only way to restore them to a dry state was to empty the stuffing out on to a canvas sheet on the next sunny day, and after constant turning over it would eventually be fit to replace. Most of the pillow bags in our little camp showed the evidence in rows of two-centimetre-long "homeward bound" stitches as a result.

As a further refinement, and throwing to the winds the problem of space for carrying, we began to think that, if we wanted to take the trouble to handle them every day, thin mattresses could smooth out to a degree the larger of the rocky lumps in our desert beds. Larger canvas squares were needed so that the complete bundle, including blankets and pillow, could be rolled for transport, and these were ordered from the canvas worker at Woomera. The amount of room taken up by these great bed-rolls was phenomenal, but we persevered, and eventually rough but strong wooden shelves were added to our vehicle bumper bars to carry them. The wisdom of our determination to include these with the sleeping gear was the subject of serious doubt on one gloomy sprinkling morning after a night of constant heavy rain. The swirling waters had discovered a minute entrance into Scotty's canvas cocoon and unbeknown to him had been trickling into his wondrous mattress all night long. When he finally emerged from the completely enclosed depths of his boudoir after somehow realising that a new day had begun, we heard him mumble something about "thirty gallons" to no-one in particular.

A year of living like this had passed before the idea of a new and devastating innovation crept into our minds. The colossal bulk of a mattress only served to cover the ant-beds and rocks so they could be lain on with more ease; so, what if we were to sleep several centimetres *above* the ant-beds and rocks? The sensation of sleeping on beds had become completely foreign to us, but our thoughts were turning in that direction and the prospect was decidedly attractive. No more dusty blankets embedded with burrs, no more lying on iron-hard ant-beds or stones, and no more grappling with an ungainly mattress-roll every morning. It was so positively effeminate that no member of our small party was brave enough to be the first to mention it.

At long last a heavy equipment fitter, who was with us at the outset, actually broached out loud the subject which was already uppermost in our thoughts. The adjective "heavy" not only applied to the equipment he fitted but also to himself, and over the months the uneven ground had surely made more of an impression on him than on us. As though nobody had ever thought of it, we agreed that perhaps some use could be made occasionally of the small canvas stretchers which we had noticed in the bulk store at Woomera. The very next trip "down south" saw us signing for half a dozen of these items and storing our mattresses in a shed as they "might come in handy" one day.

The first night back to the bush was one of unaccustomed excitement. The new beds were unfolded, and after a number of attempts, according to the varied mechanical ability of the boys, they were assembled and the much-diminished swag-roll placed sedately on top. There we were, up out of whatever surface that particular campsite presented, and what was more, the beds were capable of accommodating the entire assortment of contours with which a human body is made. It proved the most restful evening we'd had for years. Now that they had arrived on the scene, the beds were there to stay. Or so we thought.

Unfortunately, everything we owned and carried had, of necessity, to be of rugged construction throughout, and these stretchers were obviously designed for holiday campers. Their average use would normally be in the order of a fortnight per annum, in which case the fabric could be expected to last for years. In our case they were intended to be *idle* for only two weeks per year. However, those problems hadn't as yet cropped up.

After a dozen nights of blissful sleep a slight fraying began to be noticeable along the side rails which in themselves, although apparently quite flimsy, had, up till now, survived the heaviest of us.

Small holes followed the fraying, and it seemed sensible to start treating the beds with respect, gradually easing on to them, remaining throughout the evening as immovable as is possible after a day's work, and gingerly rolling off them in the mornings. Instinct dictated that to concentrate the strain on any one point, as for example in the act of sitting upon them, would be disastrous. Although they were only ten centimetres clear of the ground, that was already becoming preferable to our former height of zero centimetres, so our newly-acquired treasures rode in pride of place among our prized possessions and were treated accordingly.

One night, not much later, our heavy equipment fitter had his first major mishap. Ambling over from the fire with the misguided idea of going to bed, he carefully manoeuvred his weight on to the unfortunate little stretcher and reached for his blanket covering. The noise of the ripping canvas and the following thud could be heard throughout the camp, but his groans could be heard much further, as well as his matter-of-fact statement, "Down on the ground again." His was only the first of a series of identical episodes which put us all back to where we started, even before the mattress era.

One day a tree branch caught in the canopy of one of the trucks, leaving a suitable-sized rectangle dangling from it after the vehicle had gone past. I happened to be the next to pass the spot, and, taking down the "flag" by standing on the roof of the Rover, I put into operation the idea which had flashed into my mind the instant I saw the cloth flapping in the breeze. That night my stretcher, with its truck-canopy reinforcement, complete with "homeward bound" stitches strongly sewed with plumb-bob string, was the only one that kept its user off the ground.

The next innovation clearly must involve something stronger than canvas, and iron with heavy wire mesh was the only material acceptable. Thus our little group entered into the army-hospital-bed age, which remained with us for the rest of the project. With the U-shaped legs folded up, the beds could be hung on appropriate hooks welded to the sides of the trucks, but the problem of the hard mesh was soon found to be worse than the ant-beds and rocks. Back came the "handy" mattresses with their bulk, and at last we all had our sleeping problems solved. My bed hung from hooks looped over the bush-bashing framework on the Rover, with its bottom rail strapped to an iron D bolted to the panelling to stop it flapping. As a station boy once said, it did look as if I was carrying a gate around with me.

Of course, during the long forward reconnaissance trips I was back to a canvas square on the ground. The bed would never last long on the side of the Rover as I plunged through the dense scrub, and apart from this its iron mass would greatly affect the oil-bath compass mounted in the cabin. So on these occasions it was placed under a mulga tree and all my surplus gear not needed on the surveys was heaped on it to wait until I returned.

As we had penetrated further into unbroken country 1000 kilometres from the nearest habitation, our initial complement of one truck and two Land Rovers proved to be hopelessly inadequate. Apart from the bedding, the rations and bulk fuel drums for the machinery just wouldn't fit. Another truck was added to our retinue, and not long afterwards a third truck was found to be essential for ferrying supplies to keep us going. The complete number of three big trucks, two Land Rovers, a bulldozer, and a grader made up our camp for many years—not including the assortment of trailers.

On one "end-of-the-year" visit to civilisation Scotty and

Doug even asked about the feasibility of adding a caravan to our string. Many of these were available at Woomera, and so the following year saw us on our way with a gleaming three-berth model in tow. When it arrived in our camp (if it survived the 2000-kilometre trip), the plan was for the grader, being the only vehicle capable of dragging it over the new roads, to lug it from camp to camp. On larger sand-ridge crossings the bulldozer was always there to pull it over when the grader wheels lost their grip. The two plant operators and the cherry picker would be using the caravan, as Paul, Quinny, and I spurned the idea of being caged in at night and preferred to carry on as usual.

The 1400-litre water-tank and the 450-litre auxiliary petrol tank installed behind the cabin of each truck gave a large flat area on their combined tops, and this was as close as Quinny ever came to a bedroom. Paul always slept out in the open with only the dome of stars over him, even when they were replaced with ominous-looking rain clouds. He did this for the entire eight years, except for one or two exceptionally wild nights, one of which had occurred at a camp fifty kilometres on our way west from Sandy Blight Junction.

I had, up to this time, remained faithful to the shorter version of the Land Rover, thinking it could force its way through some of the thicker scrub by turning in smaller circles. This left me out in the elements with Paul. On this particular October night even I was beginning to think about other methods of sleeping, and as my Rover was due for a replacement anyway, I determined that my new vehicle would be of the long wheel-base variety. This would make it possible to actually sleep indoors, although I would have to overcome my repulsion of cages, and put up with larger turning circles. As it eventuated, my new vehicle went with equal ease over any surface anywhere the others would go, and at last I could obviate the need for the never-ending rolling up of a swag every day. But it had taken this particular night in question to turn the tables.

When we had started work that day a stiff wind had sprung up, and this increased gradually in force by the hour. At first fine dust filled the air, then came heavier dirt and sand followed by ironstone gravel which peppered our little group with the violence of a shot-gun and remained unabated during the daylight hours. Black clouds appeared about midday, and one or two spots of rain spattered on to the windscreens to combine with the dust into droplets of mud. The wind ripped through the mulgas and spinifex with such frenzy that complete trees were uprooted, and those already bulldozed aside were spun around and carried bodily in the direction the gale was travelling. Anything not firmly anchored down didn't stand a chance of remaining where it was, and even the big trucks rocked on their springs.

It was impossible to do anything but seek refuge from the storm by crouching in the cabins of the vehicles. The three caravan occupants swayed about in their cubicle so dangerously that I wondered if it would stay upright on its wheels, so we moved it to a position end-on to the blast. The noise of the gravel hammering on to the aluminium walls of the van was deafening when you were on the inside, but the few minutes I spent there were very welcome. We decided against trying to prepare anything to eat but each made a small meal from our assortment of tinned food while crouching wherever we could out of this fury.

When it was time to turn in, Quinny even roped down the canvas canopy flap next to his perch on the petrol and water tanks, and as the fitter had a similar position on his truck he followed suit. Paul rolled out his swag among the ration boxes in the back of his vehicle, for once leaving the iron bed right where it was on the hooks outside, after first turning his "trolley," as he called it, side-on to the battering wind.

The elements had gone completely mad already, but now a further catastrophe descended upon us. The few sizeable drops of rain were slowly being augmented by others, and by what appeared to be late afternoon a wall of water had enveloped everything. The rain was torrential, and within seconds rivers formed in our wheeltracks and boiled along the ground to the lower areas, with spray being whipped up by the wind as they went. Without reference to watches only a guess could be made as to the time of day. It looked like being a very long night.

Having the smallest vehicle in the camp, I was the only one without any shelter apart from canvas squares, but I was determined that I would not invade the inner sanctums of any of the others. My old routine would carry on regardless. However, I did think I might reinforce the one camp-sheet with a second which I carried in case it might "come in handy." If it didn't come in handy that night, then it never would.

I usually slept with my head towards the back wheel of the Rover as a sort of protection against weather in general. There wasn't much else to be protected from. Often the pad tracks of dingoes searching for food were plain to see in the dust around my swag, but as silently as they came, they crept away, never molesting me in the slightest. I had often woken to notice one or two in the flesh staring motionlessly at my lone camp, only moving when I stirred — a situation which sometimes caused me to sleep alongside my rifle. Nothing else to cause concern lived in these outback deserts apart from scorpions and centipedes, and these were only around after or during rain.

The wheel kept the bright light from the full moon out of my eyes on several nights a month and the very early morning sun in summer months away from me until I deigned to "balance on one end." Each morning this end-of-the-year sun brought with it not only soaring temperatures but also the accompanying black mist of flies. If I slept on the eastern side of the wheel during those months they would ensure my premature rising by swarming into my eyes, as the intense early heat caused me to discard my blanket.

During the winter, the later daybreak and the sun's warming rays were more than welcome after a freezing night, so I never slept west of the wheel. The most awkward combination of sun, moon, winter, and summer was consistently found to be on wintry evenings when the moon was full. As the sun set, the huge silver orb of the moon would appear in the east, sparkling down on the bush through the crisp dewy dusk and remaining to illuminate the scene almost to daylight intensity throughout the night. Trying to sleep with this phosphorescent floodlight on the exposed eastern side of the wheel burning into my eyes until the shadow of the vehicle took over at midnight was a problem I never solved before the advent of the long wheel-base Land Rover. I either fell asleep in the shade on the western aspect and froze away from the sun in the morning, or I stayed fitfully awake half the night to welcome the warmth of a new day. I usually decided on the latter course, with its added advantage of a sun-dried swag to roll up in place of an otherwise dew-soaked one.

The arrival of an evening wind added to the complication of deciding on a sleeping orientation because, in the case of wind, the feet end of the turned-under canvas swag must point into the direction from which it came, otherwise covers would be either blown away completely or impregnated with the accompanying dust and saltbush prickles or spinifex tops. If the wind was coming from the south on a summer night with a full moon, then the worst evening's sleep could be expected. If I lay on the south side of the vehicle, the last rays from the blistering sun as it set would continue to play on the already heated swag, there would be no shadow from the moon all night, and the first boiling shimmer would resume at dawn.

One way and another the science of merely stopping for an overnight camp developed into an exacting art only coped with after years of trial and error, with the error predominating.

Where could I sleep, therefore, on this memorable October night? With this unprecedented situation I decided to discard all the scientifically-accumulated rules, so I forced open the door of the Rover and waded around to the iron bed hanging on its hooks. Extreme tiredness dictated that I couldn't huddle upright in the driver's seat all night long. I quickly had the bed in place against the wheel, but by the time the canvas swag was unrolled on its wire top I was thoroughly drenched. Out came the handy extra square of canvas. I had thoughtfully spliced rope ties to each corner of it, and after battling with it in the wind I managed to fasten the corners securely to the underside legs of the frame, all but one corner which I raised a few centimetres to tie on to the iron D on the panelling.

With my boots pushed under the mudguard and on top of the wheel, both to prevent them from floating away in the night and to halt the procession of centipedes from invading them, as I had so often found before during rain, I levered myself into the structure. The saturated shirt and shorts had to go and were wedged on top of the boots in a sodden lump before I managed to complete the operation and finally install myself in the tube. I gained consolation from the fact that I would not suffer from moonstroke or sunstroke if morning ever arrived again, and there were certainly no banks of flies. In some ways I was really very well off.

The howling wind and driving rain hammered on my canvas with such uncontrollable energy that I could feel them on my body as though they were blows from a threshing machine. Another thing in my favour was that because it was October, the temperature was not actually freezing, even though the wind-blown rain caused a decided crispness to be felt.

No sooner had I encased myself, thankfully up off the swirling ground waters at least, than I was shaken to see through the slight crack in the canvas flap that the surrounding bush, camp, and country were suddenly illuminated to bright daylight. Momentarily the deluge of rain could be clearly seen as it beat on to the surfaces of everything about, and then, seconds after the pitch blackness returned, an ear-shattering explosion jolted its way through the night.

After that inaugural and savage eruption, the lightning and thunder swelled in intensity with violence surpassing any manmade nuclear device, and rampaged through the heavens for hours to come. Forked lightning laced the sky preceding crashes of thunder until thoughts of the splintered, blazing tree which had been struck near my camp eight years before leapt into my brain, and I began to doubt that I would ever see another sunrise. The metal of the bed and vehicle could attract such electrical forces but I hoped that the water cascading down over it all would serve as a conductor if necessary. At the same time I hoped everyone else was surviving safely.

Somehow and at some time during the night I must have succumbed to the effect of the day's battering and fallen into a merciful coma, finally emerging back to consciousness at a time when I could see my surroundings without the aid of lightning flashes. There was not much indication that the weather was abating apart from the fact that the bursts of light and explosions of thunder were missing. Wind and rain were still in much evidence, but, having had enough of the clammy blankets, I retrieved my hobs and clothes from under the mudguard and roped down the flap of the camp-sheet.

My appearance in the caravan reflected in the expressions of the three caged but wonderfully dry occupants as they looked at me with apprehension. Some people after an evening such as that might have become somewhat disconsolate, tending to suppress a smile, but I was so pleased it was over that I couldn't feel at all dejected. I did, however, venture a comment that the weather seemed to have the appearance of being a trifle boisterous.

That was by far the wildest night I've ever endured out in the open in the bush, but it was not necessarily the most uncomfortable. Lashing dust storms filling eyes, ears, nose, and blankets with powdery red dust, and nights of insidious invasions by billions of black specks of ants into swag-rolls made other occasions more nightmarish, but one thing was certain. The memory of that October camp will remain with me as clear as it was on that morning for the rest of my life. Even now, when I see a night storm raging outside any shelter I might be lucky enough to be in at the time, I never fail to register a feeling of gratitude.

Doug summed it up later by remarking, after a discussion of the event, that he "didn't get a wink of sleep all night anyway worrying about me." Only one day later saw me at my theodolite carrying out a sun observation in preparation for an astrofix the following night. The purpose of this was to establish a sign for the crossing of the Western Australian border where we left the Northern Territory behind. It was surely an incredible country.

Quite a while before this, Quinny, who would sometimes even beat Paul up out of bed in the mornings before the faint glow of an approaching day could be detected, made a discovery. He informed the camp of it when we all rose. It concerned the rather unusual way in which our cook dressed himself before preparing breakfast. It seemed that Paul would stand on his head to pull on his old charcoal-blackened trousers, even before he hit the ground. A long time ago he had met with an accident which had broken his legs, but apart from a difficulty in walking it hadn't really affected him. Consequently he had obviously found it easier to allow gravity to aid him in donning his clothes, and with feet in the air, the trousers could slide down to his waist. When the operation was reversed, his shirt would react accordingly.

This morning, as we camped with our road train at the Davenport Hills, I happened to rise for once even before both of them and for the first time witnessed this phenomenon for myself. Old Paul had apparently seen the glow of the breakfast fire I had lit, so although it was still dark he decided to get up and, not having his customary ration track as a screen, began to dress.

We would soon be on our slow way with our towing operation, but not before Paul balanced on one end, even though it was the wrong one.

Chapter Ten

THE BORDER CROSSING

After pulling away from this most pleasant campsite with its profusion of memories, and with everything once more on the move, my thoughts could turn to the next point of interest that this Sandy Blight Junction Road had to offer. The making of it had been so crowded with events and new discoveries from the enormous amount of exploration involved that every time I travelled over it, a sensation of anticipation accompanied me.

Of all the roads that we constructed, this had already become one which never seemed to lose its excitement, especially for me, and I think I could safely include the rest of the party in that feeling.

After once again sedately emerging from the valley in a flowing curve, we drove slowly along the straight section joining the two extremities of the sand-ridges. With less than fifty kilometres to go to reach the State border of Western Australia and our elaborate delineation of it across the road, we would soon be out of the Northern Territory. This then was the next target to look forward to. The clumsy road train trundling along behind the bulldozer would eventually negotiate the curve around the point of the sand-ridge on the southern edge of the scrubby flat to point west. Ever since we left Jupiter Well 430 kilometres back we had been travelling easterly and southerly, and this was the turning point.

We had crossed the border already because the road of necessity had to skirt the salt Lake Macdonald situated astride the border, and as Mount Leisler was to its east, that is where our road lay also. I had carried out a series of star observations just after that memorable wild night, and established a signpost to mark the point we had passed with the convoy north of the huge saltpan. Now we were about to recross that same meridian beyond which somebody had decided that Western Australia should cease.

If a huge knife was able to slice the world exactly in half, then the edge of each hemisphere along the cut line could be described as a great circle, and as each meridian of longitude passes through the north and south poles, there is an infinite number of such great circles. The equator of latitude $0°$ is the only latitude line capable of halving the earth's spheroid. All the others only pare diminishing portions off the whole, as if cutting an onion into rings. The longitude value of the north-south great circle dividing Western Australia from the Northern Territory is $129°$ east of Greenwich, and from my day-to-day plotting of our new road progress, I knew roughly when we would be making the first bulldozer cut over this invisible line.

As we had battled our way northwards, weaving through the maze of sand-ridges, I was endeavouring daily to ease the course of the road into the Northern Territory and past Mount Leisler. The Bonython Range from which I had first sighted the mountain is also situated almost on the border, south of the expanse of salt comprising Lake Macdonald. High sand-ridges barred the way for me to make the road close to the range and finally, as previously related, I became entrenched in an east-west groove only half a dozen kilometres from it. I knew then that we were approaching the actual State border-line and that any day from then we would be crossing it at right angles into the Northern Territory.

On a survey trip ahead of the road's end, and at a scaled distance from my plot, I was able to obtain a very rough idea of where the border should be. The whole valley appeared to be ideally situated for an imposing sign to advise future travellers which State they were in, for it was lined with open spinifex and the occasional attractive desert oak. The absence of scrubby undergrowth gave the whole aspect an uninterrupted view, on either side, of the parallel sand-ridges from which in this area there was so far no escape. That point of escape finally came at the present position of our train as we rounded the bend to face west.

As we now slowly drove back, I remembered the initial decision and the preparations which resulted in our finally marking this border crossing, leaving behind a monument which I hoped would survive for many years. At that time, almost half a year before, I thought of how I had stood at my theodolite observing stars in the freezing cold June nights, gathering enough readings to be able to locate the exact spot. The extra work would not be wasted on just a point of interest but would be of practical use as well, for it would establish in a visual way on the ground the boundary of the Aboriginal reserve, a feature needed for subsequent entry. Any future exploration for such substances as oil and minerals would have to be negotiated with whichever State government was concerned and this would at least indicate across our new access road where the line was. In addition, the exact latitude value of the crossing would as usual be passed on to cartographers so that they could pinpoint the location on all the latest maps.

Apart from all that, we just plain felt like embellishing this particular road with as many features as we could, for somehow I felt it would eventually be used more and more by travellers, owing to the nature of the country it traversed. Its geographical location would also make it a valuable link in years to come, and I wanted it to become a sort of showpiece or scenic drive. It has actually been in even more use than I imagined at the time, and I have never regretted all the extra work we put into it. Rains, sand-drifts, and wind have since all played havoc with the surface as we left it, but that surface is incidental to the navigational importance of the route. This is established forever, as are the desert oaks and the mountains. If travellers can stay with it, then it will surely guide them safely to their destination through some of the most memorable aspects of Central Australia.

Many times since this road appeared on maps and became known, adventurers and travellers have come to me bubbling with enthusiasm after returning from a journey over it. They have been so grateful to us for having made their trip possible that their obvious pleasure far outweighed all the personal effort and punishment we subjected ourselves to for so long to see it through to a successful conclusion.

A lot of work still had to be done before I could fix the position of the border exactly.

If the latitude and longitude of any point on the earth's surface is known, it is possible within limits to locate that point on the ground. The method is to physically fix a spot as near as possible to the precise point and then to calculate the difference between them in bearing and distance. By measuring the distance in the direction from the fixed point, the position of the required one can be obtained. The accuracy with which this pre-determined spot can be established is dependent upon several factors. One reason for a doubtful reading is the "deviation of the plumbline." This is usually an unknown in practical field observations, but it can sometimes be discovered to a degree by a series of readings using expensive and sensitive gravity-meters.

Distorted refraction conditions at the moment of observation vary with temperature and barometric pressure, but are a very real cause of consternation as they can alter by the hour. By careful selection of the stars to be used, with regard to their position in the sky and the times they are observed, these errors of celestial refraction values can be minimised. If the line of sight from the theodolite to the star is bent to cause a star in the east to appear too low, a star in the west will be similarly affected. The two variations from the mean tend to cancel themselves out, providing the two stars are of the same altitude and are observed as close in time to each other as possible.

Instrumental error is another consideration, but this can be almost eliminated by the way in which the theodolite (in this case) is used. If the optical micrometers are at variance or the telescope crosswires are out of adjustment, both "faces" of the instrument are used, and the differences, which can only be measured optically anyway, are nullified. Of course, as in everything else which is done by mere humans, personal error is yet another to be reckoned with and this is done by sheer weight of the number of readings, which share the individual misgivings and help average the results.

The Northern Territory–Western Australian border, being of 129° east longitude, now became the known value to be located on the ground. After bypassing the Bonython Range I knew it was close by. We made the road easterly in the valley to a point which, according to my plotting, must be in the next State, and it was there that I carried out the first set of observations. A relatively rough, quick fix was all that was necessary to give me an idea, to within about a quarter of a kilometre, as to how far away the border was and on which side of it I was. These results informed me that we had already crossed over and were into the Northern Territory for a distance of one-and-a-half kilometres.

The following day we carried on with the road as usual, and the next night was the one on which I planned to observe the most accurate latitude and longitude astrofix that it was possible for me to do in the bush. The site for this was going to be 1.4 kilometres by speedo reading back along the road from our earlier camp where I had read the stars the night before. This should put me within measuring distance from the astronomical station to the actual border, wherever it was. In these latitudes every second of arc in geographical co-ordinates is equal to about thirty metres, so hopefully I should be within a dozen seconds of my goal.

At noon I drove back to my intended astronomical station to obtain a latitude from the sun, put in a peg, and blaze a tree with a chiselled mark to act as a referring object. This would be needed so that I could include an azimuth or bearing observation to enable us to place several posts to mark out the direction of the "great circle" across the road.

The night was as clear as crystal, with the stars as brilliant as they could be in spite of the almost full moon which rose not long after nightfall. I was glad of the complete absence of wind, but the air was like ice. Positioning the Rover in a convenient direction so as not to obstruct the sky, I lowered the tailboard and set out the field books to record the readings. As usual I couldn't afford the luxury of a fire, because the heat shimmer rising from it would make it impossible to read the pinpoints of stars. Also, on such dewy evenings, the readings and times had to be written in soft pencil as on the wet pages ink would blur into unintelligible smudges.

Mostly I carried out this nightly work alone, and this was certainly the case now, as I thought of the rest of the camp huddled around a cheerful fire over eleven kilometres away at the present head of the road.

The object of star observations in general is to read to them precise angles of altitude and bearing and record the exact time at which each reading was made. With a knowledge of the star's angle and the time at which it was in that position, it is possible to work out when it would cross the observer's meridian. By the use of current nautical almanacs, the time can be then calculated when a transit would also be made over the zero meridian at Greenwich, and the difference between the times, converted to angle, is simply the longitude. My theodolite would give the angles, and the stop watches combined with radio time signals would supply the times.

Many variations of these basic principles are brought in, depending on such conditions as partly clouded nights, the time available to prepare a programme, and the sort of result required. Latitude values can be obtained without any knowledge of time at all, based on vertical angles only. All that is needed is to apply observed angles to the star when it is at its highest point or culmination (which is actually at transit over the observer's north-south line) to its angular distance from the equator. The distance from the infinite production of the earth's equator to the star is its declination. This application can easily result in the knowledge of the angle of the infinite production of the earth's axis to the horizontal which is automatically the latitude. There's really nothing to it.

This was all in my mind as I set up that freezing instrument in the spinifex valley between the sand-ridges and began to point the telescope skywards.

A torch shining on the mark at the blazed tree gave me enough light to relate a set of readings from the stars, and by midnight I had pages of balanced observations ready to be reduced under much warmer conditions than these. Half a dozen stars in the east paired with an equal number in the western sky would reveal the longitude; and a similar number in the north and south would yield my latitude. Others at random would supply a bearing accurate enough to enable the placing of the marks along the border when it was finally located.

At last I could pack away the instruments, books, and lamp and return along our brand-new road to the camp to waste the rest of the night in my swag. Once there I couldn't resist a few minutes by the still glowing embers of the campfire, thawing out my stiffened hands.

When the new day arrived it brought with it a high wind, making the scene very cold indeed, but while some maintenance was being done on the bulldozer, I began the sums. The excitement of discovering just where the border lay goaded me on until late afternoon when I triumphantly drew a line with a flourish under the three hard-won results: the latitude and longitude of the astronomical station and the true bearing from it to the blazed tree. By applying all the known corrections, and from inspection of all the values which gave the final mean result, I thought the position should be within a close tolerance of seconds of arc. I was particularly elated to note that this astronomical station was only four seconds of angle west of the border. This meant that at the rate of thirty metres per second, we had only 120 metres to measure easterly to establish it.

The valley in close proximity was especially free of obstacles, and I was sure our boundary mark was going to prove an easy-to-see, useful, and interesting feature on the road. I could already visualise future travellers stopping at it to discover that they were passing from one State to another, after reading the information I was about to stamp out on the usual aluminium plate.

The following day, being Sunday and a day of so-called rest from our rigorous routine, gave me an opportunity of returning to the scene of the future activity armed with my results. I measured the calculated distance with a tape and drove in a bush-wood peg I had cut for the job, but the procedure of lining up others along the direction of the border would have to wait until I brought the rest of the camp back to help. As I had already imagined after finishing the sums, the line fell in a most ideal spot. I decided to use the next few hours on the axe.

A large straight desert oak stood about a hundred metres away on a sandy slope, and, reluctant as I always was to touch such beautiful specimens, I drove over to it. A long length of its hard round trunk would make an ideal post to sink alongside our new windrow.

These trees, like almost all others in these harsh conditions, are as hard as iron, and it took a good deal of axework not only to chop it down, but to separate the shaft from the branches. Even though this area was so remote that I'm sure it had never before been visited by white man, I nevertheless felt sad when

Top and Bottom: Some of the fascinating Aboriginal rock carvings I discovered in a creek-bed near the Sir Frederick Range. Those in the top photograph have been temporarily outlined in white for easier identification

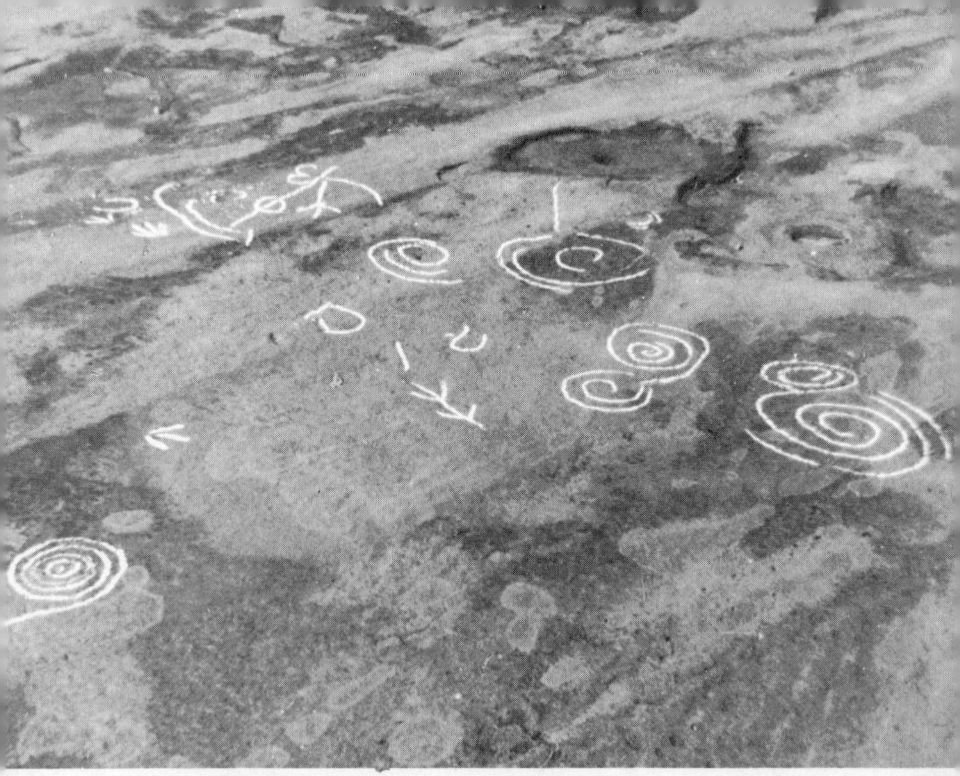

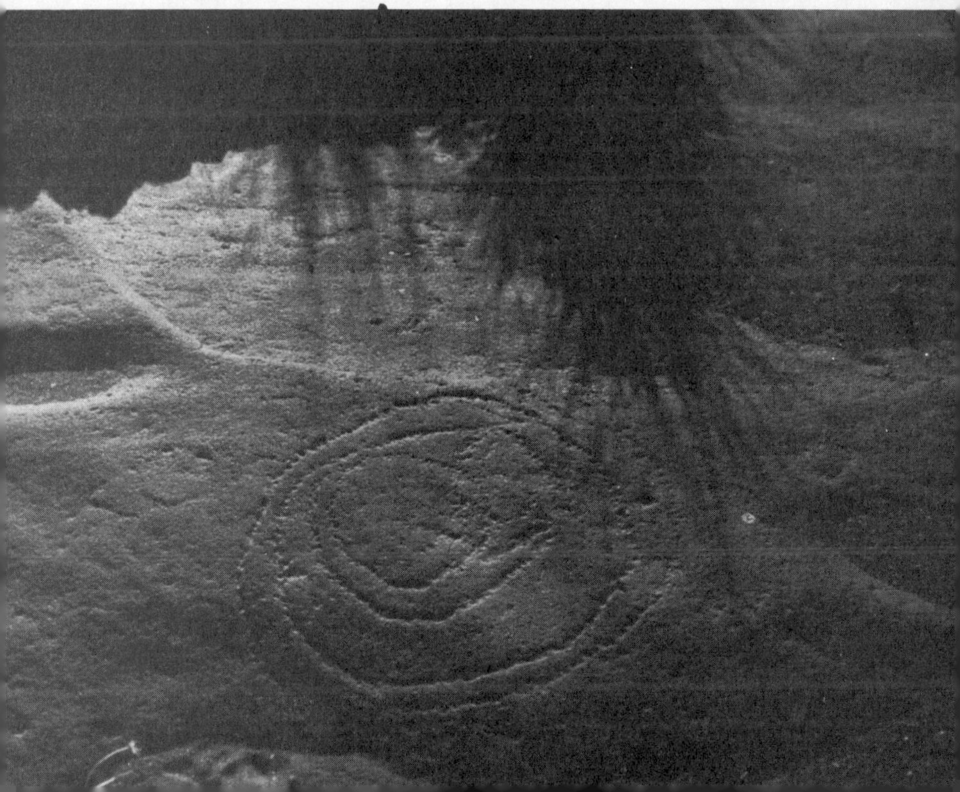

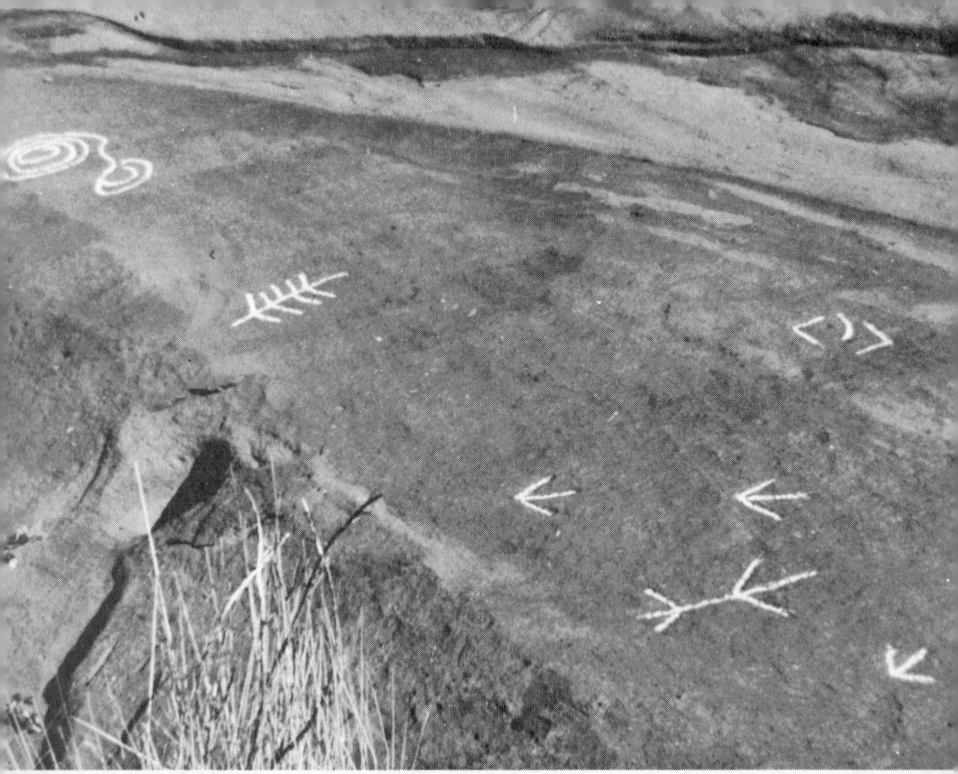

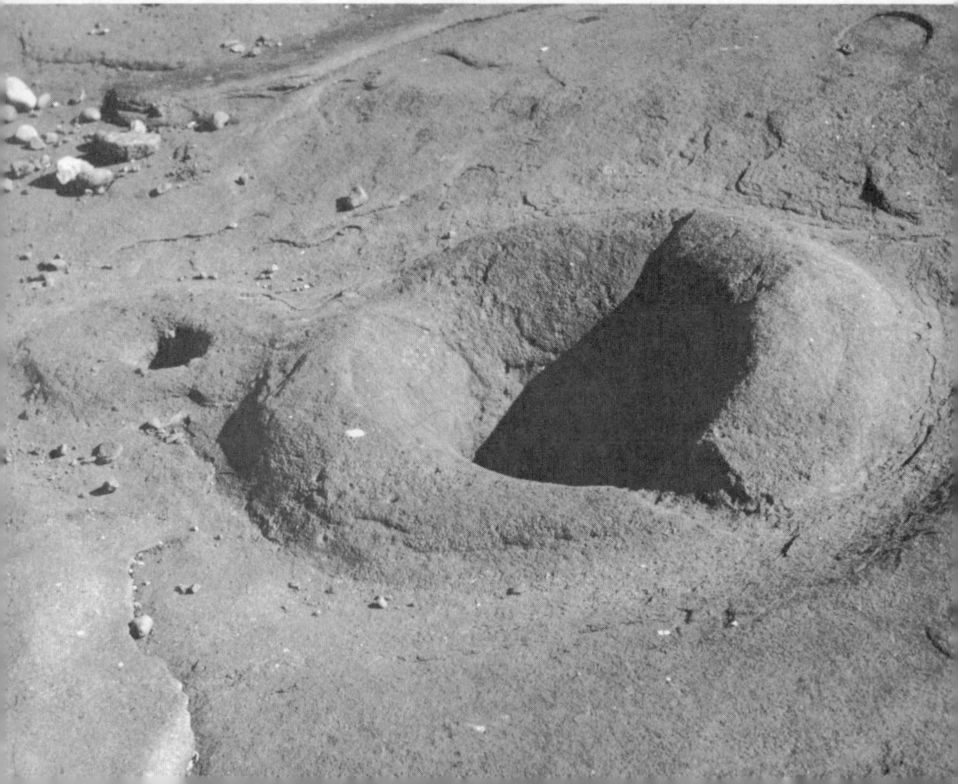

the stately oak toppled. That it would be put to good use didn't compensate fully for my sense of remorse, but there it was on the sand and a lot more work was to be done to prepare it for its final resting place. Three others would be needed to define the meridian, but the boys could obtain one each from the grove not far distant from this one, and in this way the guilt could be shared.

After much chopping with the sharp axe, I found I was unable to move the trunk to the road on my own, so my work on this quiet sunny "day of rest" being done, I drove back to the camp.

The next day saw us all back at the border laying out the line across the road which if produced would, or rather should, arrive at our other border-crossing and sign fifty-eight kilometres true north of this one. Then again, if produced to the south, it would cut another of our crossings 267 kilometres away, and yet another border signpost I had established on our road in the Great Victoria Desert 536 kilometres distant, before eventually converging on the south pole. This direction was obtained from the azimuth observations I had made to the blazed tree reference mark.

While Scotty, Doug, and Eric faced up to similar trees of their choice with their axes, Quinny and Rex helped me move my huge log over to the road. We dug a metre-deep hole in the ground alongside the peg and levered the post into it. As I had spent what was left of the previous afternoon stamping out the relevant information on the aluminium plate with alphabet punches and hammer I had it ready, and I now attached it with the rust-proof brads to the squared-off portion facing the road. We placed this post on the southern side of the road, thus allowing the sun to shine forever upon its gleaming surface throughout each day. Every time I am shown a close-up photograph or colour slide of it by an exuberant tourist I have been pleased with my illuminated theory.

The lead capping came next, and by the time the signpost had been rammed solid with the end of Paul's crowbar the rest of the boys were helping each other to load their logs on to the workshop Rover's bonnet.

Eventually a second post was inserted in the ground across

TOP: Further examples of the rock carvings. Such work is extremely rare in this part of the country. BOTTOM: A natural sink hole found near the carvings. Aborigines had chipped a gutter into the circumference to admit trickling rainwater

the road, followed by a third one hundred metres to the north, and lastly a fourth the same distance to the south. The four were lined up exactly with my theodolite and capped with lead sheeting. The ever-present tin of white paint came last, and then the row of freshly-painted white posts was in place forever.

I climbed up on to the slopes of the sand-ridge on the northern edge of the valley and, with the bright sun behind me, finished the job off with a carefully composed photograph of it all. Scotty, having the only other camera in the camp, followed suit, and we were all ready to return to our camp where the bulldozer and grader were waiting for us. A day of such activity, full of interest and achievement although by no means restful, was as good as a holiday from our never-ending ritual of roadmaking.

On the following day, we all resumed our work of changing the face of Australia, causing every published map in existence to become out of date, with renewed eagerness. Quinny was due to leave on his normal trip back to Giles for supplies so while we bulldozed ahead, he drove off in the opposite direction. As we saw him go, we thought how the ration truck would be the first normal vehicle to pass between the pair of posts astride the new road and over the magic, invisible line of the border.

A decade later, when I made a trip along this road, I saw that, sure enough, the Department of Aboriginal Affairs had made use of this location. A large sign advising travellers that they were about to enter an Aboriginal reserve, and setting out conditions for entry, stood alongside the road close to the post bearing the aluminium plate.

The latitude of this crossing appeared on the sign I had made resulting from the astrofix with its value of $23°41'43''$ south, and of course the longitude quoted was simply $129°00'00''$ east, the known value of the border. I had punched these into the soft metal underneath a straight vertical line cut with a cold chisel, with NORTHERN TERRITORY on the left of it and WESTERN AUSTRALIA on the right. The date of the astrofix and crossing also informed anyone interested that all this had taken place on 10 June 1960.

Keeping the sign to its simplest form, the foregoing came under a heading across the top, adding emphasis to the fact that this was really THE STATE BORDER. I bravely signed the bottom so that surveyors in the future would know who to blame if they had cause from the ensuing geodetic surveys to disagree with my efforts. I had a reasonable idea that by the time this happened, if at all, I would be in a place where their harangue could not reach me.

Quinny and I in advance of the train now pulled up at our "gateway" to await the arrival of the rest of the camp, and I used the time to repaint the posts from my handy tin kept upright on the passenger's floor of my Rover. Frank, who had crossed and recrossed the spot on numerous ration and supply trips during the months that it had taken to make the road ahead, merely switched off his engine and read his cowboy book. Normally we would sit and yarn over some happening or aspect of our work as the bulldozer dragged the bulk of the camp towards us, but with me vigorously wielding a paint brush he could see that this was out of the question on the present occasion.

Not having passed between the exalted posts since that cold June Sunday of activity, I think Doug might have cast a second interested glance as he jolted along, as probably would Rex and Eric. Scotty would be wearing an expression of satisfaction at a job well done, but I doubted if his passenger in the workshop Rover would even have stirred. Old Paul without his ration truck would probably be fast asleep.

Chapter Eleven

THE DISCOVERY OF A DESERT TRIBE

Even if only a few kilometres over the border, the camp that night was once again in Western Australia and within sight of the Bonython Range. Only the upper half of the rocky outcrop of which the range is composed could be seen, protruding above the high sand-ridge to the north. We had been paralleling that same wall of sand from well back into the Northern Territory, as it didn't recognise such a thing as a State boundary. It had been the factor governing the course the road was made to take as we had battled our way northward through the maze of sandhills during construction. This insurmountable barrier reared up to protect the range from invasion as efficiently as a moat protects a castle.

This overnight camp was at the turning point in the road where I had finally conceded a victory to the sandhills by turning easterly with the survey. I remembered endeavouring to cross over it on the original reconnaissance, having deflated all the tyres to a point where any further slackness would have rolled them off the rims. After a dozen futile attempts, a lower saddle to the east had finally allowed the charging Land Rover admittance to the rocky range beyond, and I could at least climb to its beckoning summit. From the high vantage point I hoped to scan the skyline in the direction we wanted the road to take and thus help in its final location.

I knew that we would not be putting the new access over this gigantic mound of sand because, even if we did, very few future travellers could have used it. The sand would have surely blown back to cover the tracks, leaving an access open only to specially-equipped vehicles.

The effort involved in eventually reaching the highest point of the range, with the last steep pinch being only negotiable on foot, paid dividends of paramount importance. It offered me that first exciting glimpse of the huge blue bulk of Mount Leisler far off to the north-east and made me decide then and there to which immediate destination the road was bound.

Months before, when we had arrived at this turning point with the bulldozer, I had abandoned the Rover on my "expedition wheeltracks" and climbed up on to the heavy machine with Doug. We were going to test whether the high sandhill could halt a dozer with its great steel Caterpillar tracks. We soon found that even these wouldn't pull the massive weight up the sharp incline ahead, as there was a complete absence of grip in the soft sand. A spot slightly to the west looked more feasible, and after much churning and manoeuvring, the multi-tonne giant sat balancing on the top of the ridge. That this was the first such event in the history of the sandhill and indeed for this area in general since the world began was indisputable.

We stood on the engine cowling, which was rumbling with the steady beat of the pulsating motor beneath, to view the scene spread out in front of us. Voicing the sentiments for both of us as we beheld the stark wastes of dead trees scattering the otherwise bare sand, Doug uttered the first description of it ever made by white man: "Gosh! what a desolate joint!"

On that first survey trip, I had covered 200 kilometres return from our camp at the current head of the road only fifty kilometres away. Weaving and circling through that dreaded belt of sandhills searching for a route to make the road accounted for the hundred extra kilometres of exploratory bush-bashing. A comment I had made in my little diary at the end of the expedition stated clearly that there were millions of them, a figure which I suspect was grossly inaccurate but summed up my feelings at the time. At the close of each day I would sit on my opened swag by the Rover and briefly record the happenings on that date. Although some of the information such as the foregoing could not stand up to scrutiny, the general atmosphere would be obvious on later reading.

One day, when I was the furthest distance from my camp on that particular survey, I remembered seeing volumes of black smoke spiralling into the sky ahead coming from somewhere beyond the Bonython Range. This smoke, which continued to rise throughout the afternoon, would have originated from Aborigines' fires as they burnt off large areas of spinifex during their hunt for food. Small lizards inhabit the bases of the spiky hummocks as in summer months they are the only source of shade, and the sharp needles of the dreadful *Triodia* also protect them from eagles and any other form of danger, with the exception, of course, of human beings.

These lizards are only about fifteen centimetres from head to tail and dart with lightning speed from clump to clump. As we drove over seemingly endless plains of spinifex with bare red sand separating each mound, we found that these energetic creatures were forever part of the scene. Some would dart across the metre-long expanses of clear sand with their lower legs only, the front legs and the majority of their little bodies in a vertical position. As their rear legs churned over the ground with supersonic speed the action so closely resembled that of a cyclist that the logical name of "bicycle lizards" resulted. They are in evidence throughout the entire year in all seasons, surviving the longest droughts, and as such they provide a constant source of food for Aborigines.

Some Aborigines can actually catch these creatures by hand, but to obtain them in sufficient numbers for eating, they merely light the spinifex up-wind and as the fire spreads the lizards are laid bare. Enormous areas of the resin-loaded hummocks are persistently burnt all over Central Australia by the few scattered Aboriginal tribes, and the billowing clouds of dense smoke are a characteristic betrayal of their presence. From long experience, I have concluded that these aerial indications could originate at a distance of anything from a kilometre or less to fifty. In one particularly remote area in the Gibson Desert we saw great smoke activity at a distance which seemed to be only a sandhill or two removed from us. After beating over sandhills in that direction for over twenty kilometres in a Land Rover, with the smoke still appearing to be the same arm's length away, we gave up and returned to camp before darkness fell.

On this day of reconnaissance, as I became entangled in the convolution of piles of sand, I noticed that the smoke columns kept erupting from an area in exactly the direction of my planned destination, so I discarded the compass in favour of the signals and simply made for them. By the time I reached the final mountainous wall of sand, with the rocks of the range well in view on the other side, I was surprised that I hadn't yet caught up with the actual fires. Nevertheless, when the sun had set, taking with it the last glimmer of daylight, and intense darkness had settled over my lone camp in the quiet bush, I sensed I must be quite close to my unseen neighbours.

No sound came to me through the stillness of the crisp night air, and the blackness of the near-by range added to the eerie gloom surrounding me, but I could feel the nearness of other human beings. Having come across so many of their camps, I could see in my imagination the naked shapes of the Aborigines huddled close to their microscopic campfires, extending a hand now and then to feel the warmth. Nothing but an occasional whisper would pass between them as one by one they would lie down among the spinifex on the frosty dew-soaked sand to simulate what their pampered white brothers would call sleep.

As I put my little diary back into the leather survey bag which was my constant companion in the Rover, and crawled into the canvas-covered swag-roll, my thoughts dwelt totally upon those freezing Aborigines. How near to me were they in this limitless expanse of wilderness, and had they ever set eyes on a white man with a motor vehicle? Never in this part of the country: of that I was quite certain. We were 500 kilometres in a straight line to the nearest settlement of any kind, and that in itself nestled in a very remote locality. Nothing whatsoever would be found for twice that distance to the west or south.

When I pulled the cold canvas over myself I found that it was already soaking wet, and as I lapsed into oblivion after the long hard day of crashing through the scrub, my last reflection was how soft we had become. Remove the camp-sheet, take away the blankets and raggy clothes, and I would be the same as the Aborigines, lying out in this frigid atmosphere without protection of any kind. They didn't even have the contents of a tin of bully beef in their stomachs as I had, but only some scraps of a lizard or so which they might have been lucky enough to burn out of hiding. They would certainly have heard the engine as I charged over the sandhills and would know of my presence as surely as I knew of theirs. Their keen sense of hearing would have told them a foreigner was at hand, and how far away he was, with much greater accuracy then I was capable of when I tried to guess at the distance of the smoke.

I couldn't help the predominant feeling that I might have visitors before the time came for me to move on in the morning. After dark the Aborigines generally remained at their fires, but I had experienced incidents where definite evidence of their silent infiltration into my camps could be discovered. A bag of flour ripped open or an empty water-tank on my Rover with the tap full on over a patch of wet mud were the sort of things I had previously encountered. Of course, the loss of the water gave the greatest concern, and I only hoped that the same thing was not about to happen this night. Taking a length of fencing wire and a pair of pliers to the unprotected rear of the vehicle, I took the precaution of clinching a strong loop over the handle and under the outlet pipe in such a way as to render it immovable by hand, before finally encasing myself in the swag. As always, the water provided my only means of returning to the far-away camp, leaving the food to rank a quite insignificant second. Radiators just won't work on tinned meat.

Awake at first light next morning, I threw back the stiff, frosty canvas, being anxious to get a warming fire going as quickly as possible. Although I had not actually lain shivering throughout the night I was by no means over-warm, and the frozen open-top hobnailed boots into which my bare feet were thrust didn't fail, as usual, to reduce the temperature even further. In less than a minute, the match dropped into the nearest clump of spinifex had converted it into a roaring inferno on to which I deposited an armful of debris from the bonnet of the Rover. The accumulation of sticks from dry mulga trees and bushes as I pushed through the scrub would heap up daily until it resembled an eagle's nest, but I always left it intact for use as morning wood, except when it piled higher than my line of sight through the windscreen.

Only after I had stood by the fire for a few minutes did I thaw enough to inspect the sand around my airy camp to see if any nocturnal visitations had occurred. However, the crisp white frost lay all about quite undisturbed. There was no need even to look at the water tap, but when something so vital to survival is at stake, I did cast a quick glance in that direction before hurrying back to the source of warmth. The strong wire clinch was in place. As I stood soaking in the warmth of the flames I looked to the north past the rocky range. Sure enough, the hunting fires were still issuing their columns of smoke. If the terrain in that direction had been at all favourable I would definitely have attempted to make contact with the hunters, but as it was I had already pushed the little vehicle as far north as it would go — a fact proved by the way I had had to close the final gap to the summit of this range on foot.

There had been much to do that day, as I had far from discovered a satisfactory route to make the road up to this point. The normal procedure I had evolved was to endeavour to sort it out on my return trip. With this in mind I rolled up that soggy swag and with no thought of breakfast started up the engine once more. I cut the wire safety device from the tap with my pliers and huddled over the dying embers of the fire for a last warm-up before climbing into the refrigerated cabin. I had already wiped a rag over the icy windscreen in order to see through it, and was soon moving slowly over the hummocks, tossing like a cork on a rough sea, in the general direction of camp. I had been pleased to notice the absence of any flat tyres which so often opened the day's proceedings, thanks to the lack of hard dead mulga roots and branches. These had given way to more open spinifex since the last session of patching the day before.

As it was early winter at that time the sun quickly dispelled the biting cold of the evening, and by mid-morning I was warm enough to discard my old army overcoat as I battled with the steering wheel. The engine helped to defrost the cabin, and with an occasional glance behind, I could see the columns of smoke becoming less and less distinct as the morning wore on. I was never to see smoke in that direction again, and I frequently wondered who the people were, how many there were of them, and how far they actually were from my lonely camp in the sandhills.

Two years later, after the completion of that road, an incident happened which caused me immediately to remember those smoke spirals and to wonder if in fact I had at last actually made contact with one of those mysterious hunters.

We had joined our road at Jupiter Well with another coming down from the opposite direction and originating from Gary Junction, the northern terminus of the Gary Highway. Having thus paved the way for a 500 kilometre loop access back to the Gunbarrel Highway, it became necessary to drive on past the burnt-out truck to Giles via the Sandy Blight Junction Road for more supplies. We intended then to connect Gary Junction to Australia's Number One Highway near the Indian Ocean, a distance of some 500 kilometres further away to the north-west.

During the drive from Jupiter Well I had Scotty with me in the Rover. We had left his grader in readiness at Gary Junction to be guarded by Paul, Eric, and Doug at our camp there. As usual they were quite content to await our return. Quinny of course came with us, bringing his supply truck to be loaded with fresh rations and fuel for our next lunge into the Great Sandy Desert.

As we drove past the burnt shell of our old ration truck and approached the border signpost, we were actually only about sixty kilometres due north of our former camp near the Bonython Range. The distance removed from my original estimation of the location of the hunters' smoke fires was therefore only forty kilometres, which isn't very far when you are speaking of nomadic Aborigines in search of food.

We both saw it together. Away to the south and from behind

a bank of high sandhills erupted a long, thin column of black smoke in the exact direction of the rocky range where I had camped alone with a strong wire clinch on the Rover's water tap. The events of that night leapt into my mind. The coincidence of the proximity of both sightings, although seen from vastly removed standpoints and with the space of two years separating them, was far too strong to ignore.

Grinding along the still relatively new road, I repeated the story of that last experience to Scotty. No sooner had I finished than, as if in answer to our combined curiosity, a naked black figure appeared from the far side of the nearest sand-ridge, running as fast as it could go. We had been wishing for time enough to negotiate the sandhills in the direction of the smoke so that we could try to bring about a meeting with these people, but now that looked to be unnecessary.

There was a clearing on the crest of sand several hundred metres from us, with scattered scrub and low bushes covering the intervening distance. It was in this clearing that we first saw the movement, and it was obvious that the Aboriginal was racing in a line towards the road in such a direction that he would be able to head us off several hundred metres in front. No Aborigines had ever been seen in this locality in the years we had been working there. As we closed the distance between the Rover and the running figure we caught an occasional glimpse of the athletic form as it flashed between the bushes. Then, in almost no time, our eagerly searching eyes beheld the wiry frame of an Aboriginal standing on the road ahead, barring our way. I was surprised at his boldness, as in most such instances these people are quite timid in their approach.

He waited until we stopped, and we could see no evidence of his having just raced at least half a kilometre. No panting or gasping for breath, not a sign of tiredness as he strode around to my door, and no obvious traces of sweating. Had an untrained white man extended himself to such a degree he would have been slumped down with his head between his knees, fighting for air and soaked in perspiration. We were soon to discover that although visible evidence of exertion was lacking, this didn't comprise the whole description of our audacious friend.

I began to open the door as the wild-looking Aboriginal reached it, blocking the action, and instantaneously the space between us through the open window was converted into a dense seething mass of flies. For a time I could barely see him through the haze from which emitted such a din as to drown the sound of the engine. Each individual fly must have been screaming at full pitch and the noise was deafening, but this wasn't even half of the disruption to our former comparatively serene drive. Forcing its way through the fog came a fragrance with such impact as to cause my passenger, anxious as he was to view our visitor at close range, to vacate his seat forthwith. The operation as he rocketed out of the cabin was performed with such downright alacrity that I was scarcely aware of being alone in so microscopic a period of time.

I was a metre closer to the source of the aroma, and this didn't help a bit. Furthermore, I was trapped in the confined space of the driver's seat, and any attempt at retreat was foiled by the radio transceiver box installed in place of the middle seat. Forward evacuation was thwarted by the door which, although slightly ajar, was held by what easily could have been a hand, obscured completely by the flies. When the time came to breathe inwards once more, after exhaling for only half a minute, I just had to push open the door enough to permit my escape.

Admittedly the day had been warm, but as there hadn't been much rain for about a year and with the bathing facilities as scant as they were in the desert, the likelihood of even a personal rinse since the last good downpour was extremely remote.

Not a word in either of our languages had as yet been uttered, as conversation necessitated periodic inhalations which, as well as the scented atmosphere, would have drawn in a hatful of flies with every breath. The Aboriginal, still gripping the door to save himself being knocked over as I plunged out into the open air, thrust the swarm of insects on his shoulders in through the window, obviously in search of anything edible. How he could see beyond his eyelids eluded me. At last, standing ten metres away over the windrow of dead spinifex, I could regain my breath. It seemed reasonable to give him a tin of water from the tap at the back of the Rover, and as I turned it on, I thought of the wire twitch with which I had locked that very tap two years before.

Could this have been one of those mysterious hunters against whom I had guarded my meagre but precious supply of water on that frosty night at the Bonython Range? Would he have been the one who might have crept silently into my camp and caused my failure to regain the party at the head of the road? After all, it was only forty kilometres or so from where we stood at that moment. My own answer to these questions was that definitely, here in the flesh, in the midst of that mound of flies, an original member of that same tribe was drinking the proffered water from that self-same tap.

After several futile attempts to discover the number of his companions still concealed behind the sand-ridge, we indicated that we were about to make our departure and leave him to return to his people, taking with him his personal cloud of flies, quaint perfume, and all. Even after the flies had receded into the background and we had resumed our drive, it wasn't until we reached the campsite that the last haunting traces of that unforgettable stench had finally dissipated and it again became possible to breathe *inside* the cabin.

Right now, all this was in the future. My momentous meeting with one of those remote nomadic hunters was still to come, and at this time I could only cogitate over their existence. As I lay on my swag in the intense heat of the Central Australian summer, only the memories of those earlier smoke sightings and the reasons leading up to the enforced change in the direction of our road filled my mind. Complete quietness had replaced the former screeching and banging of our bulldozer train, and it now stood on the road alongside our little camp, waiting to resume its towing activities in the morning.

But on every subsequent trip past this point on the road, following that first meeting with the aromatic tribesman, I have always recalled that it was about here that another happening well worthy of mention had taken place.

Chapter Twelve

THE SIR FREDERICK RANGE

In the bush the most reliable alarm-clock during the summer months is provided by the flies. Therefore, living up to their well-earned reputation, they informed us promptly at five o'clock in the morning that the time had arrived to be considering some sort of movement. Paul already had his fire going and was handing back goodnatured insults to Quinny as fast as he received them with reference to his recent dressing operation while standing on his head.

The caravan which trundled along as last carriage in our train had long since become utterly unfit for occupancy. The dust on the road, already as fine as flour from being churned up by the bulldozer in the lead, became increasingly voluminous as it combined with the clouds created by the crippled grader, second in line. The big four-wheeled water trailer attached to the grader was almost lost to view altogether as the addition of both lots descended on it and mixed with its own billowing quota. Then the complete combination of the three enveloped the unfortunate travelling boudoir. Nothing whatever could be seen of it at all, for it was completely obscured by red, powdery bulldust throughout the whole time that we were mobile each day.

The vacuum effect of the aluminium cubicle had been steadily sucking in the atomised laterite since we began this long haul, and once, when we prised open the door, we found the interior to be quite unrecognisable. What were formerly beds had been transformed into hills of dirt gently moulding on to what must have been the floor thirty centimetres below the visible surface. A miniature landslide had occurred as soon as the door was opened and the built-up mound against it was suddenly released, and the whole aspect resembled the interior of a cave.

Added to the somewhat dusty appearance, the interior of the caravan had also taken on the temperature of a blast furnace with the bottled-up heat from the merciless ball of fire in the sky. The unrelenting rays beating down on it for a dozen hours a day had combined with the dirt to force the usual inhabitants to live as they used to, out under the stars.

The flies by no means confined their activities to a favoured few but paid equal attention to everyone, and in no time we were all balancing on one end and ready to be on the move as soon as possible. Once back on the dozer Doug was completely free of the insects, but they continued to plague the rest of us until the sun sank below the western sandhills once more. Some schoolboy once wrote that "the sun sets in the west, then hurries around to the east to be in time to rise in the morning," and at this time of the year that "hurry" seemed to be supersonic.

With every kilometre we were to travel that day in our new direction south, we would be moving further away from the centre of Australia. The nearest point on land to the Timor Sea was just 1000 kilometres due north of us, and we were 900 kilometres from the Great Australian Bight in the opposite direction. The closest the Indian Ocean came to us was also 900 kilometres away to the north-west of this overnight camp, with any other sea being over twice that distance in other directions. I often thought of the position of Sandy Blight Junction as a geographical coincidence, being as it is exactly equidistant from the bodies of water named. A circle with a diameter of 1900 kilometres can be described, with that junction as its centre, and be wholly contained by land. Outside the circumference the seas are touched tangentially at three points 1000 and 1700 kilometres apart from each other.

TOP: Bulldozing the second cut on the road to the summit of the Sir Frederick Range.
BOTTOM: My Rover on a sandhill near a specimen of *Gyrostemon ramulosus*, the dreaded camel poison bush

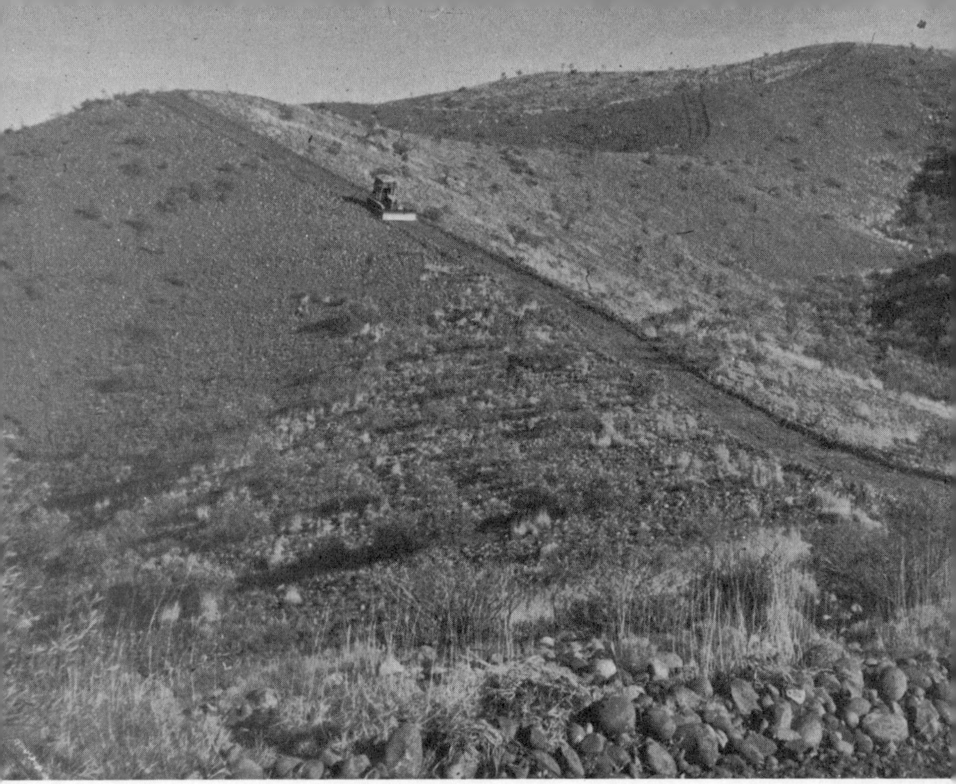

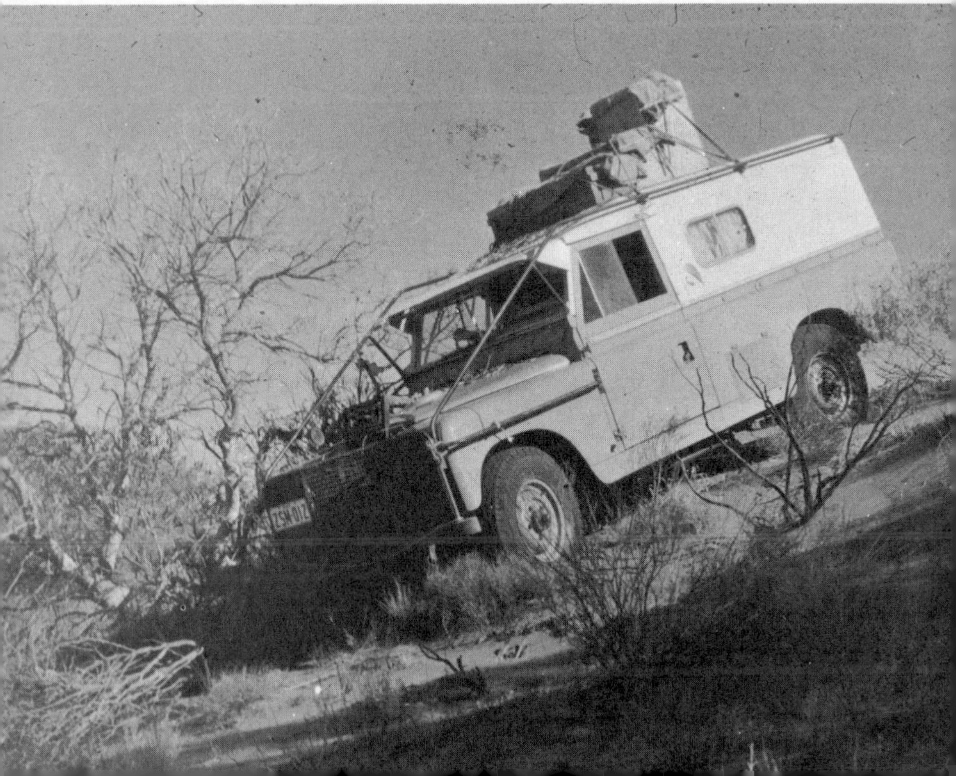

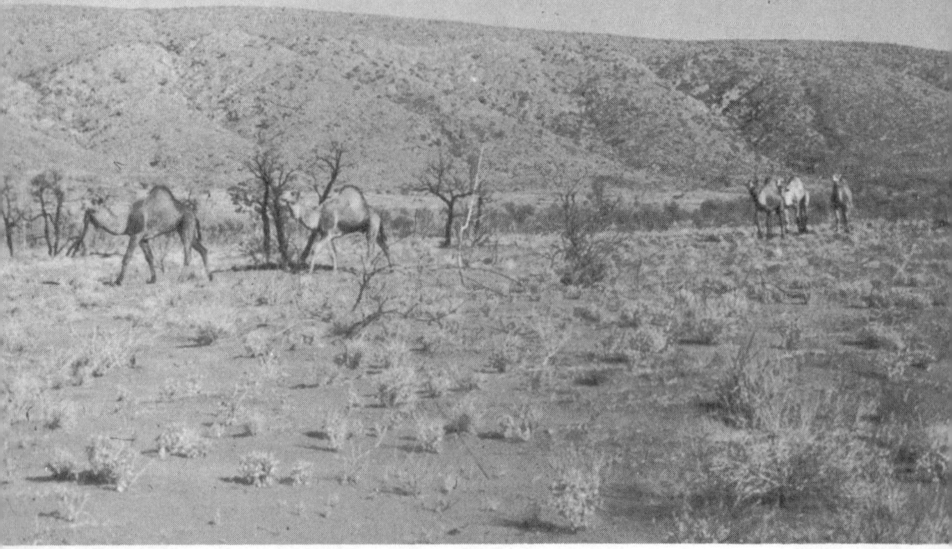

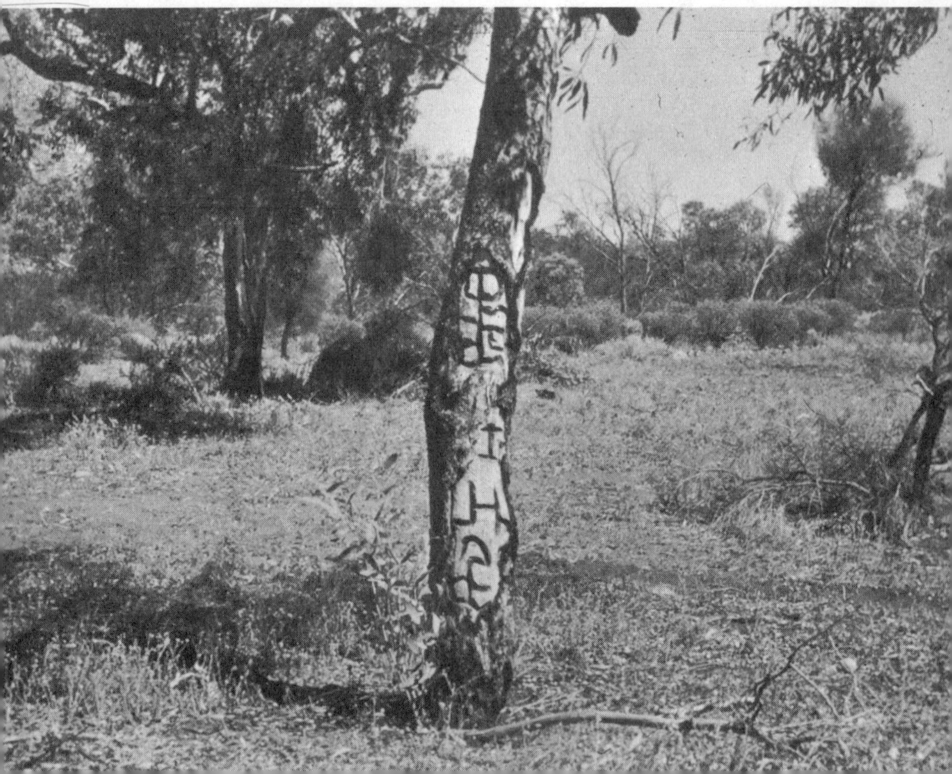

We had been widening the gap from the north at a snail's pace since leaving Sandy Blight Junction, then maintaining an even distance after the Davenport Hills, but now we had turned south once again. I had been looking forward to this day as eagerly as I had to all the others, because yet another exciting feature was about to be revisited.

The Sir Frederick Range was still about sixty kilometres away, putting it well beyond the scope of a day's towing distance, but somehow I always experienced a sense of anticipation whenever I approached it from either direction. The vast collection of massive bare convolutions of which the range is composed appear quite differently from every aspect depending on the time of day. Early morning sunshine highlights the eastern faces of each fold while throwing deep shadows into the separating gullies, giving a stereoscopic effect. With the late afternoon's rays the picture is changed completely, presenting to the traveller an entirely different pattern of light and shade, and the whole concept varies between these extremes throughout the day.

The surfaces of the rounded hills are paved with orange-sized pebbles among saltbush, and each crest is joined to the next by narrow saddles, making it possible, in the right sort of vehicle, to drive to the highest point. As this summit was only 210 metres lower than Mount Leisler, the establishment of a future survey beacon upon it would be a certainty. When I first visited this range I determined immediately to ascertain whether or not a vehicle could reach the crest. If it could, a bulldozed spur off the main access would greatly help the follow-up survey parties from the National Mapping Council. Heavy equipment was needed at each trig station to carry forward the survey and, where time permitted and if it was at all feasible, I would endeavour to pave their way with a road. Mountains such as Leisler were out of the question for such an operation, and heavy batteries, tellurometers, theodolites, and tripods would have to be carried on their backs all the way.

In this case it had appeared from first impressions that we might be able to construct such an access, and as our camp had been only a dozen kilometres back, I decided to try and find a route. Some of the grades would obviously be quite steep and if

TOP: A herd of wild camels near the Walter James Range. BOTTOM: We found this bloodwood tree covered with Aboriginal carvings near Bungabiddy

something dire happened in my attempts, I would have to cover the distance for help on foot. That is, of course, if disaster only struck the Rover and not myself as well.

With the road and camp to that date having reached a point within sight of the Sir Frederick Range, I had set off on a relatively short reconnaissance in order to plan the best way of closing the distance. The closer the main road could be made to skirt the foothills, the shorter would be the access spur to the top, if of course one could be done at all. In any event the road's proximity to the hills would make the future survey either easier or more difficult.

With all this in mind I left the camp early on the morning of Friday 13 May 1960.

Throwing superstition to the winds, and successfully negotiating the intervening dry creeks and rocky forerunners to the main range, I had soon found myself at the point where the first of the bare hills reared up from the flat surrounding plain. With the radiator pointed at the first stony rise in front of me I carefully selected all the strongest gears the Rover had to offer, and with its wheels all contributing their share to the traction, the brave little vehicle began the ascent.

Small round pebbles were flung in all directions as the tyres competed against each other for grip, and soon the only thing to be seen through the windscreen was the blue sky. Some of the stones were much larger than others, and these had to be bypassed, but eventually that first minor crest had been won and I could see a razorback linkage on to the next higher one. It was just wide enough for the wheels, and with the edges sloping away on either side at an impossible angle, great care had to be exercised to prevent one half of the Rover from slipping over.

All the while I was looking at the route with an eye to the great bulldozer blade flattening out the sharp edge of the saddle which allowed the access across to the next hill. After crawling upward over ever-increasing slopes for two-and-a-half kilometres I could at last begin to think that a helpful access road could be put into effect. Already the view to the south stretched away to nearly a hundred kilometres, spanning the country over which we had been making the road ever since leaving the previous mountain range, now plainly visible.

Then came the last pinch which would put myself and the Rover on the highest point of the range. There had been many doubtful parts of the ascent that could have prevented the whole idea from being carried out, but this final slope looked like being the deciding factor. Being almost at an angle of forty degrees, and with the loose stones still prevalent, I really didn't think I had a hope. The angle was all right but the pebbles reduced the traction to almost zero, and once I attempted the climb I knew I would have to keep going or risk the possibility of tumbling hundreds of metres down into the adjoining ravine. But nothing was ever gained by too much weighing of the consequences, as I had found so many times in the past, so, leaving the Rover door open for a quick getaway, I began the adventure.

As I fought to keep the long axis of the vehicle aimed at the summit, the four wheels attacked the problem of moving it forward and upward. The stones flew in all directions, but I sensed that we were gradually winning. On top at last, I dared to look back down over the crest to the valley below. I still doubted the dozer's ability to follow this route and thought what a spectacle it would be to see its fifteen tonnes spiralling end over end down into the canyon.

The faint blue outline of the last range from which we had just made the new road could be seen eighty kilometres away due south, and parts of the Petermann Range were visible on the south-eastern skyline. Although hidden by intervening sand-ridges, a large irregularly-patterned salt lake had forced an otherwise straight route from the previous range to be dog-legged away to the west. This last set of hills, the Walter James Range, was 100 kilometres away.

The present outcrop was named after Sir Frederick. Just who he was I hadn't been able to ascertain, but he certainly had a much more impressive looking topographical feature named for him than I did. My private mountain, Mount Beadell, was 380 kilometres away to the south-west and could not be seen from here, but nevertheless I was grateful for small mercies. I was content in the knowledge that it could be easily seen at a distance of 380 metres, disregarding the relatively insignificant prefix "kilo."

The view from the top of this range gave a feeling of complete freedom from a world of rush and bustle and congestion, I thought, as I climbed out of the still intact Rover—the first vehicle to be in such a location since the mountain had been created. As I scanned the horizon and immediate surroundings, a second spur which radiated from the crest became apparent in the general direction our future road would be taking. This one looked far more possible as an access than the way I had just come. Almost anything would have looked better to me after that last hair-raising incline, and even though it was still very steep I decided I would prefer to make my descent down this other route.

Now that the object of this day's reconnaissance had been fulfilled, I re-entered the weary Rover and headed for the immediate edge towards the new spur. Going down proved, as always, much less exciting than travelling skywards, because for one thing the hazards were visible in front of me and not behind, where they could be left to the imagination. Half slipping and rolling over the pebbles and half driving, I reached the first rounded globule which was linked to the next and lower one to the left by a saddle. After that, each successive hump could be reached by means of these narrow connecting necks. It was all much safer than my earlier ascent, and by the time I had reached half way I realised it was all over and the road to the top had been located. I wondered about the bulldozer on the final stretch to the top, but even if only that were made accessible by vehicle, the survey parties following would have their task made infinitely easier. We would soon find out.

At the halfway point, my present course brought me to the only area on the whole incline large enough to turn a vehicle around, and I knew that we would have to level sufficient room with the dozer to enable such a manoeuvre to be carried out. If a traveller attempted the first half and doubted the condition of either his vehicle or his nerves to complete the journey, the alternative would be open to him to turn around after walking the rest. I would erect a sign on the main road at the foot of the range to that effect.

Back in the camp, I excitedly informed the boys of our intended operations, and by the time I had finished I felt that they were actually looking forward to the challenge. Doug told Scotty that his dozer would sail up to the top but he doubted if the grader could make it. Scotty confidently informed Doug that he would patiently wait on top with his grader for Doug, if and when he managed to get there with the dozer. This theme went on into the cold night as we sat around our cheery fire until we turned into our swags, anxious to get on with it in the morning. I must have been equally eager, as I leaped up out into the freezing morning air and smashed my watch glass in my hurry to get to Paul's breakfast fire. I used to hang it on a nail on a fold-up chair alongside my pillow, to be able to ascertain the time using one eye and without disturbing the blankets. As I jumped out at first light, I slammed shut the chair in the same operation which caught the watch between the arms in a pincer action, shattering the cold glass into a thousand fragments.

A day later we had closed the distance with the new road from that camp to the foot of the range and gained the spot where my second set of wheeltracks had finished up the day before. We were then ready for the onslaught to the top. I had marked the previous camp with a neat blaze on a fine big desert oak and an aluminium plate giving the latitude and longitude values I had observed from the stars and the date, also the distances either way to the next places of civilisation. Of course the forward distance, as yet unmade, would have to be added later. I had painted the notice after attaching it to the tree with galvanised roofing nails and brads, but even these had failed to save it from souvenir hunters later on. I drove past this tree years later to find only an empty blaze and a memory of what useful information I had left for future travellers. For the benefit of anyone who happens along the Sandy Blight Junction road and notices this vacant blaze, the information they would have read would have told them they were in latitude 24°03′56″ and longitude 128°39′09″, with a distance back to the Giles weather station of 126 miles. Now I would need to have quoted it as being 203 kilometres. The date on this plate was stamped 17 May 1960.

Just past the point where we would be making the T junction of the offshoot road, we bulldozed a square in the scrub to which the camp could move during our attempt at the new "scenic drive" construction. I drove along my latest wheeltracks with the great bulldozer in hot, if slow, pursuit and we arrived at the halfway point without incident. Opening out the ledge to the maximum area to enable a turn-around platform for the grader in the event of its failing to reach the summit, we pressed on and up with the heavier machine. The only thing really preventing most vehicles from negotiating the grades were the looser surface stones, and once these were removed by the big steel blade, it should be possible with any four-wheel-drive conveyance. But could the diesel engine drag that enormous bulk up the rest of the way?

On and on it crawled, over minor pinnacles and connecting saddles, until the final test was at hand. Not only did the shiny Caterpillar tracks effortlessly pull the mass of steel to the peak without so much as one falter, but they pushed a huge bladeful of rocks up in front as well. It was a magnificent feat for such an ungainly machine. There it stood on the highest part of the Sir

Frederick Range, on the exact site of the future trig station, as visual proof of just how powerful the engine really was. The work that had preceded this operation of putting a bulldozer up there with a road stretching back from it to civilisation in the dim distance was immeasurable.

As we stood on the cowling of the engine and saw the road snaking and writhing back down the mountainside like the track of a big dipper at a fun park, we noticed the grader sadly backing away to the turn-around point. The rubber tandem drive tyres, big as they were, could not get the necessary grip to raise its weight up the slopes. A deal of friendly exchanges would pass between the operators that night.

Our task completed, the dozer blade was once again lowered. A second layer of stones would be shaved off the road on the way down, making future trips by mere motor vehicles all that much easier but nonetheless exciting. Back at the bottom we made the junction with the main access road more noticeable in the scrubby area it happened to be, and I erected a sign on a post I had chopped in readiness. The distance to the road terminus at the trig site was three-and-a-half kilometres, and so our preparatory job would surely save future surveyors a lot of hard work.

With that behind us, the need had arisen for a further reconnaissance to determine the location of the next stage of the road, so the very next day I plunged off once again into the scrub ahead.

Still thinking about the success of our venture of the day before, I soon broke out of the scrub, after collecting another staked tyre, and made for a visible opening between two sandhills ahead among more desert oaks. On the way the line caused me to cross over several dry watercourses which originated from the foothills of the main range, and as I crossed over one, a blackish area in the creek-bed caught my eye a dozen metres off to the right. It appeared at first to be sand. This had probably been carried down by previous rains which had set the creek flowing temporarily and had been deposited as the water swirled around a bend, but the colour of it was intriguing. I had never seen anything like it before.

I stopped on the far bank of the creek-bed and ambled back

casually to satisfy my curiosity. As I drew closer I was fascinated to discover that the smooth surface was actually solid rock, and I was soon clattering my hobnailed boots over it. Suddenly, a mark on the pavement at my feet caused my pulse to quicken, and on darting glances around a wider field I saw more and more of them. Aboriginal rock carvings in this sort of country! Finger paintings with ochre on cave walls and roofs we had seen, and even tree carvings, but these were actual chippings of mysterious tribal patterns into solid rock! Would the surprises that this area had been shielding by virtue of its extreme remoteness never end?

Neatly carved spirals, concentric circles, shapes of animal and bird tracks, and geometrical designs were revealing themselves for the first time in history to the eyes of a white man. Their age could only be guessed at. As I gazed at this display of primitive art I tried to imagine the scene which must have taken place so long before. Aborigines must have sat on this very rock within sight of the range now known as Sir Frederick's and now with a bulldozed road up to its peak, quietly chipping out their designs with heaven only knows what sort of implements.

With the excitement and anticipation of discovery known only to those who have been in similar circumstances, I searched far and wide for further evidence. However, this bare expanse of smooth black rock was the only one I came upon. By a sheer stroke of fortune, I had happened to cross over this diminutive creek-bed within sight of it. Even so, had it not seemed out-of-the-ordinary in colour and appearance, I probably wouldn't have given it a second look. I have since brought a professor of anthropology to this place and he was positively ecstatic at the sight, photographing the hieroglyphics from every angle for further intensive study back at his university.

It was certainly a discovery of major importance and one which added another vital link to the chain of known data from which an overall picture of ancient habitation in Central Australia is constantly being compiled.

All this, then, was the next goal to look forward to as our

cavalcade drew away from our overnight camp at the Bonython Range. With our lethargic progress we would not reach that area that day, but would have to content ourselves with settling down for the night in the midst of the sandhills on the way.

But our anticipation would surely make those sixty kilometres yet to travel pass by almost unnoticed.

Chapter Thirteen

ROADS DO MAKE IT EASIER

From a bird's-eye view, roads on the surface of the earth appear merely as thin threads seemingly meandering along to some sort of far-off destination, bypassing obstructions, curving over others, and generally avoiding anything that might impede their progress. If bulldozed and graded with microscopic equipment, the path of an ant as it weaves its way through grass and around sticks and stones would be comparable to that which actually takes place, magnified millions of times, in a roadmaking project. Substitute a Rover for the ant and mountains for the stones, and clear the debris from the wake of the scout ahead, and the result will be an accessible road for all to follow. Everybody knows this and realises that somebody must have planned a route in the first place, but how many people speeding along highways or scenic mountain drives pause to give even a second thought as to why the bend ahead curves the way it does or the reason behind the direction of the previous straight stretch? The answer would be almost 100 per cent—nobody.

If the road curved right instead of left fifty kilometres back, then the traveller would not be anywhere near his current position, and the whole of the road would have been vastly different. Or suppose that, after he had travelled 500 kilometres along a highway and had a similar distance ahead to his destination, the whole road were to be suddenly erased completely. Where would he go? In what direction did his destination lie? How could he negotiate the maze of lakes and rocky outcrops, hills and swamps in front, and how could he even return to his starting point? His whole life relies on a narrow access barely wider than his vehicle, and the knowledge that it will lead him to his goal by the best route possible, without his having to think about it at all. Nevertheless, in the beginning someone *did* have to think about it.

As we entered that belt of sandhills separating us from the next range after pulling out of camp that blistering November morning, I couldn't help thinking just how much work had gone into the planning and execution of making that snake-like track ahead. All we had to do now was to drive along it, turn with it when it did and be raised over hills or lowered into gullies as it carried the great weight of our train towards our still far-off destination. We had no more decisions to make as we went; whether to pass the mountain of sand in front on our right or left, or which side of the outcrop of white gypsum to make for. The road told us that, and it would do the same for any future traveller. The initial battle with this section of virgin bush had been done forever and there was no more need for astrofixes, compasses, staked tyres, and surveys. For us they would all come again, making as many demands on men and machinery as ever, but in a new area vastly removed from where we were at present serenely driving along a made road.

With every bend and change of direction, I remembered how I was forced to laboriously try every avenue open to me before selecting the one we were now on, and the weeks of lone reconnaissances which went into it all. It had been only six months before and already the road had been used by a few follow-up survey parties and Aboriginal patrol officers, and of course there had been the constant comings and goings of old Quinny on his fortnightly supply trips. I was glad that we had made it a practice many years before, at the commencement of all this work, to widen the first cut to the width of two blades of the bulldozer, thus saving the overhanging mulga branches from raking along the sides of the vehicles with every trip.

Over the eight years this project had taken, that alone would have added up to a sizeable amount of extra effort and fuel, but now as always when travelling on our roads, I was sure it had paid us back in full. Nobody would bring out a bulldozer again in that country for many a long year after we had gone, if ever, and I couldn't contemplate not doing what we did at the time.

I was quite aware that in the course of time rains would come to scour gutters in the surface we were leaving, and the slow growth of hardy desert scrub would encroach on our cleared roads. Sand-drifts would cover tracks over the dunes and ant-beds, and rocks would be exposed by raging wind storms. The smooth surface we were forever leaving behind us as we went would be battered by the elements daily. But I was also sure that those same elements could not erase where we had been, and the final path that our two huge machines had carved through the bush and over the outback of Australia would remain forever. The course beaten through the formerly untouched and unmapped country in which we were working would survive to guide people through that wilderness safely to their journey's end. They may not even be able to drive exactly on the surface we had called a road after metre-deep erosions and general disrepair had assailed it, but if they had a vehicle, an animal, or even their feet capable of keeping with it, then it would lead them to safety.

After all, even before the road was there in any shape or form, and with that remote country still as it had been since the creation of the world, a little Land Rover had driven over it three or four times already. I'm sure it was grateful that it had any sort of a road to travel on at all, and that at least temporarily it didn't need to batter its way through confusions of dense mulga trees and between thorny branches of "dead finish" bushes with their needle-like spines each scraping their own share from the outer skin of the unhappy vehicle. On a graded surface its underside was not raked from dawn to dusk by upended roots of dry trees torn out of the ground by raging August winds, hummocks of spinifex, eroded gravel, and creek-beds. Its tyres also remained actually inflated for days on end. Only grim reminders of its pioneering days manifested themselves even on the road in the form of the iron-hard, wooden daggers which would eventually work their way into the tubes to flatten them.

As we drove in the heat on the dusty, sunbaked cutting, I wondered again just what would happen to some lone traveller if, when half-way along its length, the whole road were suddenly to vanish.

This particular section through these sandhills had finished much better than some others, as it had been raining on and off while we were working on it and so the surface had compacted into a more solid mass than it did when dry. As the moist doughy mixture was pushed about by the dozer and driven over many times immediately afterwards by the heavy grader, followed by the rest of our vehicles, a hard crust had resulted and this eventually dried like cement. Normally, powdery dust is left after the initial clearing to be smoothed out with the grader, but the texture is basically unaltered.

Here and there, still in evidence off to one side or the other of the road, I could see the old wheeltracks I had made on the reconnaissance survey weaving around sticks and bushes. A trip out ahead and a return over the same tracks left an imprint which usually lasted for years. After the comparatively frail little vehicle had skirted daintily around obstructions and indicated which way to go, the massive bulldozer had a way of ironing out any unnecessary bends and leaving the original wheeltracks to coincide only occasionally with its finished straight line.

One tree-studded area 300 kilometres west of Sandy Blight Junction had proved itself to be an unusual exception: the reconnaissance tracks and the final road were one and the same. In fact, even after the road had been made the trees were still there, standing up between the windrows, unharmed even after the last pass of the grader. We had never before struck these peculiar sorts of trees, and Paul took delight in describing them to astounded and disbelieving audiences for ever afterwards. "There were only leaves; no trunks or branches—just leaves." He told them that it was no wonder I could drive in a straight line through them in any direction because, with the absence of trunks, there was nothing to stop me. Warming up to his subject, he would go on to point out that of course they had not been damaged by the bulldozer, repeating his rhetorical question: "How could they be with no trunks?" The leaves, apparently not needing support, had been left in mid-air three metres above the finished road in the same position as they had always been.

The explanation lay partly in the fact that old Paul's eyes were not as sharp as they used to be, and partly in the extraordinary properties of these most unusual trees. At first casual glance from a distance, as Paul said, only their leaves could be seen. Nothing seemed to be holding them up. I remembered wondering what I was running into when I first drove into that wide flat valley between two parallel sand-ridges. I had never seen anything like this before, and I began to wonder if perhaps I had been too long in the bush. However, as my laborious progress brought me nearer, thin smooth stems at the base of the trees became discernible. Their colour had up till then blended perfectly with the background, rendering them completely invisible. Three or four metres from the base, a measure of foliage suddenly blossomed from the top of the otherwise uninterrupted length, becoming quite apparent long before its wiry support.

My goal lay far ahead in the shape of a saddle in the southernmost sand-ridge which was gradually converging on my line of direction, and as these trees were only a little over a centimetre or so in diameter, I didn't have to veer one way or the other to miss them. Their spindly trunks easily bent over as the Rover touched them, so I kept my straight course for the saddle without thinking too much more about them. When I had discovered that we could construct the road over the lower part of the ridge I turned around and headed back for camp, but was amazed to see that all the trees over which I had driven were back in place with my tyre tracks on either side of them. They had succumbed to the vehicle when it had gently bent them on to the ground in order to pass over them, but afterwards they must have sprung back upright again.

I described the trees to the boys at the camp when I got back, jokingly omitting to mention their trunks, and on the following day the bulldozer followed my tracks with its blade barely skimming the surface. After it had passed by, all these trees were standing up again as usual between the impressions left by the steel tracks. Then along came the grader, with its blade only trimming off the surface bumps, and again there behind it, as large as life, stood the trees as if nothing had happened. Eventually the only way we could clear them off the road was with our axes, but not before Paul had arrived with his ration truck to shift camp to the end of the road.

Another spot where my original tracks and the road coincided was just over a hundred kilometres further on, again west of Sandy Blight. I had been driving over a rise, pushing through a small but dense patch of low scrub, when a blinding flash penetrated a break in the leaves not fifty metres off to the right. It was like the reflection of the sun's rays from a mirror shining directly into my eyes and, wondering whatever it could be out in this desolate country, I gave up my course and made straight for it. Half-way to where I estimated the source of the light lay, the shrubs cut out completely, giving way to an open patch of ground on a miniature bluff with the surface sloping sharply away on the far sides. The brilliant gleam had been momentary, having vanished almost as soon as I had seen it, but as yet I could find no apparent evidence of the reason for it.

I drove out into the clearing and suddenly the explanation became clear as dozens of small glassy surfaces showed themselves emerging from the ground at all angles around the Rover. Geological hammer in hand, I jumped out and began to pick at one patch which flaked off in sheets as I struck at it with the sharpened end of the implement. It was obviously mica. As I searched further around about, I could see some actual "books" extending out into the air from an underground source. Anything different from sandhills, travertine limestone, and quartzite in these areas always caused great excitement, but here was something we had never before discovered anywhere. Suddenly, without warning and without any previous indications, an outcrop of books of mica appeared as if from nowhere. So that it would serve as a point of interest from then on, I decided to make the road pass within easy viewing distance of this strange phenomenon.

With that day's survey at an end I drove back to the camp with my news. It was becoming almost monotonous the way I would return to the boys full of great enthusiasm at some exciting discovery I had made and to relate what was in store for them. It also became monotonous the lethargic way with which my news items were received. I was always prepared for their looks of disbelief which plainly told me that I was over-dramatising my descriptions, and I would have been disappointed had they reacted differently. This only served to inspire me to relate bigger and better tales of what lay ahead, while all the time I was confident that later on they would secretly agree there had been something to warrant my sensationalism.

The following day I returned to the outcrop to guide the dozer to the spot, and the road was soon carved out alongside it. With the heavy machine to replace my geological hammer, Doug aimed it at the centre and lowered the blade. The result was a huge mound composed mainly of hundreds of books of mica, and many more were exposed for the first time to see out of the sides of the miniature crater left by this operation. It looked like a vast open-air library of books, all with transparent pages. Even Scotty condescended to climb down from his grader upon arrival to search through the rubble for interesting specimens. At the same time I'm sure old Paul and Quinny still didn't know it even existed.

While waiting on a sand-ridge crossing for the convoy to catch up, I happened to notice an out-of-the-ordinary tree a dozen metres off the road. It was of a species I hadn't seen very often. Upon closer examination I remembered a story from Ernest Giles' journal which described in detail how, on a later expedition than the one on which Gibson had perished, several of his camels had been poisoned. They had eaten herbage from certain trees which had been prevalent in small patches but which were also dotted about at random along his course. Several years before, that same species of tree had been pointed out to me by a surveyor who was also an extremely knowledgeable amateur botanist. We had been together on an expedition 500 kilometres west of where I was at present, and one of these trees had been growing alongside our campfire.

There could be no doubt that this specimen, with its "goanna skin" patterned bark and light green foliage, was one of the dreaded camel poison bushes. My friend had identified it as *Gyrostemon ramulosus*, and this agreed with the name given to it by the botanist Baron von Mueller who examined specimens collected by Ernest Giles almost a century ago. He had lost one camel which had eaten the leaves, and others had barely survived from the effects, staggering about uncontrollably until they dropped. Their respiration and lungs seemed to be mainly affected, and there was profuse frothing at the nostrils, depending upon the quantity consumed. Giles had barely managed to save several of his string upon which his life, not to mention the success of his expedition, depended.

Not many other references have ever been made to this species or its effects, but more seems to be known about a close relation, namely *Gyrostemon australasious* which is also named as a camel poison bush. Horses and other animals are equally affected by the toxic feed, but if after eating it any lucky or less greedy ones happen to survive, they will avoid it forever after. Only in dire straits in the desert, when food is at a minimum, do camels have to be forcibly kept away from it. It is possible that some doubt could have arisen in the identification of the scant samples brought back by Giles.

In any event, as I waited on this hill for the others to catch up, I thought it was amazing how much more I seemed to be able to notice while travelling on the finished roads, compared with the times when we were working to make them.

There had been no shortage of cleared campsites on this slow return journey, as every half dozen kilometres or so there appeared beside the road a neat little square which we had made for overnight stops during construction. As the day was coming to a close I drove on to the next one after the train had come and crawled over this poison tree ridge to prepare a cooking fire which would be ready for Paul's arrival.

The bare patch on which we would be camping that night happened to be situated within easy walking distance of an extra high sand-ridge over which we had been forced to put the road. The ridge went on for many kilometres in either direction so we had chosen its lowest point for our crossing. It had been on this particular hill that Quinny had used all his inventive skill in trying to return to our camp near Sandy Blight from a long ration and supply trip.

On that occasion he had gone over a thousand kilometres back along the new road to pick up a huge load of diesel drums from a dump we had established for our new operation. While the bulldozer was there Doug had carved out a big T-shaped pair of ditches two metres deep with an inclined ramp into which the supply truck could be driven. Over a hundred drums of fuel had been offloaded alongside in such a way that one man could load a truckful without help by rolling them over the level tailboard. Quinny had just visited this and had managed to stack almost thirty drums on to his truck unaided. With this great weight he headed back for our camp, and he had finally arrived at this high sandhill crossing before the grossly overloaded vehicle gave up the struggle. An ordinary truck could and often had negotiated the steep rise, even if it had to make several attempts, but not so on this occasion. Quinny was there on his own with hundreds of kilometres yet to go to reach our fuel-starved machinery and not a hope of crossing over this hill.

The big vehicle was fitted with an excellent winch and long cable but there was nothing strong enough to use for an anchor. Any frail desert growth would be useless and no trees of any size were ahead within reach of his steel rope, so in usual Quinn fashion he hit on a solution. He calmly dropped his tailboard, rolled out half his load of drums, and tried again. When he found that this didn't make too much difference, he carried on with another plan. Rolling the whole dozen or so heavy drums one by one up and over the crest ahead, he finally had a great pile of them down the far slope and within reach of his cable. It took a powerful man to accomplish this feat, for when the grade was too great for him to push the drums he painstakingly levered each one up the soft sandy incline with two chopped lengths of mulga branches. That finished, he took his shovel and buried one of them three-quarters of its length into a great hole he had dug out of the sand. Then he manoeuvred all the rest to a position on the uphill side of the one he had chosen as the anchor, so that they would act as an immovable barrier against which the buried drum could push.

With his winch motor running on the other side of the hill, he dragged the long cable up and over the crest and wrapped the free end around his "post." This had all taken a great many trips back and forth, but eventually he climbed up to his cabin and engaged everything. Low reduction gears; four-wheel drive, which the truck had already been using for days; and finally the winch. The strain was applied and the anchorage over hill held solid while the big and still heavily-laden vehicle slowly crawled up to the top. Quinny didn't stop until he was on the far downhill slope and had reached his huge pile of full diesel drums. He had pulled that enormous mass of steel over simply by a different arrangement of his own load.

All that remained was to roll all these obstacles to one side of the road, exhume the last one, fill in the hole, and carry on to our camp. He had no hope of reloading them up on to the truck, but he knew that once he was back at the camp, we would return to the spot and recover them. I was sure that if it had been absolutely necessary he would have devised a bush way of getting them all up on to his load again before leaving. It was at times like these that I was particularly pleased that I had chosen the members of our little party so carefully.

His valiant truck had suffered a further setback after it had struggled on from this sandhill and landed at a point quite close to Sandy Blight. All those attempts to cross over the hill finally proved too much for one of the axles and it gave up the fight just after he had turned a corner. Endeavouring to battle on in front-wheel drive only, he soon became badly sand-bogged and had to resort to his spare axle to repair the damage. He had been carrying it for many thousands of kilometres for emergencies such as this, but when he tried to install it, none of the wheel studs would fit through the flange holes. It was the wrong axle for his vehicle. But the Quinn determination was undeterred, and he began diligently scraping at the holes with a screwdriver in an attempt to enlarge them.

That was how I had found him three days later when I went out in my Rover to search for him. During all this time he had been worrying at those holes, hoping to make at least two studs fit, but it had proved a losing battle with the high tensile steel. With the vice on my vehicle to hold the axle and a rat-tail file I always carried, we soon had several studs in place and it was no trouble for him to drive straight out of the bog with all four wheels pulling. Back in camp we transferred the load to Rex's fitter's truck, and most of the party returned with Quinny to this memorable sandhill to fetch the remainder of his original load. I used the time they were away to carry out a long reconnaissance ahead in readiness for resuming work after this slight interruption.

I had a cheery fire going for the convoy when it thundered in to pull up on the road adjacent to the cleared square. The fires we would make on this operation were for cooking only and were kept as small as possible because of the great heat of the day, so different from the frigid weather we had experienced when we were camping on this same spot before.

On that occasion we had camped on the same site for two

days while we worked on the crossing in an effort to reduce its height sufficiently to allow the many future supply trips to take place without undue interruption. Now, as we sat about in the cool of the evening between the big sand-ridge and the still glowing bulldozer train, we all agreed that we hadn't altogether succeeded.

Chapter Fourteen

LAKE HOPKINS

Within minutes of leaving the camp early next morning, the bulldozer had pulled its heavy train effortlessly over the big sand-ridge where Quinny had spent so many long hours of backbreaking work. It was lost to view immediately as it descended sharply into the adjoining spinifex valley on the other side. Meanwhile the others leisurely packed up after the overnight stop. They knew from weeks of this same procedure that they had plenty of time to catch up before the midday meal which, after all, would be consumed only a matter of twenty kilometres away.

As usual I preceded the train to roll off any branches or mulga trees which might have blown back on to the road before the oncoming weight pulverised them, making the operation much more difficult. Using the base of the tree trunk as a pivot, the winds had a habit of spinning the larger diameter end of foliage around like a top on its side, describing a perfect arc, and this was just as easy to roll away again. To prevent a monotonous repetition of the same thing I usually chopped off some branches to interrupt the flowing action until the foliage dried out. With the dead leaves lying on the ground, only dry sticks remained, and these would not so easily be caught up in the winds. There weren't as many loose trees around as there might have been because of the way in which Doug usually left them. After bulldozing them down and out of the way, he would raise the blade and drive on over them, converting them to matchwood, so the wind had nothing to push against and merely blew harmlessly over the flattened mass.

Also, by driving in front I was out of the dense volumes of dust left in the wake of the train, and I was able to return for Rex in the event of any possible mishap to any of the towing linkages. An enormous strain was constantly being applied to everything and the way it all kept coming proved what a solid job had been done on the draw-bar weldings.

In any case the whole thing had to stop every two hours as the track rollers needed a complete regreasing with the big volume gun to keep them going. The heat generated by the non-stop rotation of these comparatively diminutive wheels as they took the complete weight of the bulldozer and its screeching Caterpillar steel tracks made them such that they wouldn't have lasted without careful attention. To save Rex from lumbering along behind eating the dust to keep up with this never-ending operation, or trying to judge the distance all day in front, the volume gun was carried on the machine itself and Doug and I would do it ourselves.

When thinking of greasing anything, the average motorist has in mind a small grease gun a few centimetres in diameter and about a dozen centimetres long. The volume gun required for the bulldozer is a four-gallon affair with a great lever and handle powerful enough to force the life-saving lubricant into the various working parts. Even this seemingly clumsy task requires much care to ensure the grease is not injected at too great a pressure, for this might damage the flimsy seals which retain it inside. Once the seals are disturbed, sand, dust, and dirt can enter and cut the bearings to pieces, and if this were to happen our mammoth towing project would quickly come to a grinding halt.

This day of travelling would be in heavy sandhill country all the way, with the exception of the brief respite provided by the Sir Frederick Range and its environs. This of course would amply offset the lack of topographical features on this section of the Sandy Blight Junction Road. As we rumbled along I remembered clearly the reason behind every bend and curve over each sand-ridge and the difficulties we would otherwise have been confronted with had we not put the road where it was. The weeks of reconnaissance and spinifex bashing I had put into the original survey were still vividly imprinted in my mind's eye, and as each crossing was negotiated I felt more and more satisfied that the amount of work at the outset had paid us back in full.

Eventually, around midday, the Sir Frederick Range began to make its appearance over the tops of some of the lesser dunes, and soon we were emerging out into the clearing which was from this approach the forerunner to the black rock of Aboriginal carvings in the dry creek-bed. The two sizeable sand-ridges either side of the road petered out abruptly, giving way to open country where we could see many kilometres ahead without the persistent obstructions from sandhills. The heat, bottled up in the high ridges, seemed even more intense and would surely be again after we passed over this almost pleasant area ahead and re-entered the next equally thick belt.

I didn't resist the urge to drive on across the intervening space to the carvings and spent the time once again examining my valuable discovery of six months before. There was a good area for a dinner-camp near by, so I gathered some dry sticks for a fire to put on a billy for tea. The area wasn't too near this hallowed spot as I wanted the locality to remain as natural as it had always been for the untold centuries since the Aboriginal artists had been at work. When the rest of the party arrived, shortly after the dust had settled behind the train which had pulled up by the fire, Scotty and Doug joined me at the rock for a last look. Quinny, who had been struggling back and forth past it on the supply runs ever since we made the road, stayed with Paul as he worked in the open blazing heat at the lunch preparations. Rex took up his usual position at the diesel drums, refuelling the dozer and replacing the fuel and oil filters. Oily dust with the consistency of mud trickled off them as he took them out and piled them neatly on the fire to keep the vicinity of the carvings as clean as possible. We would be burying everything before leaving in an effort to set an example to travellers in this special place at least. Actually it would be extremely doubtful if anyone happening along would even be aware of the existence of this primitive open-air studio with its treasures, and only an expert with trained eyes might see them. That sort of person would be as eager as we were to protect them and treat them with the uttermost reverence they merited.

The dinner camp being over, Doug and I pulled out once more, and it wasn't long before we drove past the turnoff that we had bulldozed to the summit of the Sir Frederick Range. The road to the top could be seen snaking its way up over bare hills and razorback saddles, and I well remembered what it had taken to discover a route. Now, in that heat and with so far still left to go, we simply drove slowly onward, and in an hour or so the range was left behind as we were once more swallowed up in the next belt of sandhills.

Once we had negotiated this forty-kilometre stretch, the sandhill crossings would be virtually behind us for the rest of this journey. Not that the sandhills cut out by any means, but up to that point I had managed to run the road between the ridges, paralleling them for lengthy distances at a time but not actually crossing over them. It had been one of the most difficult sections to battle with, demanding hundreds of kilometres in hard reconnaissance and many astrofixes to keep us constantly trending in the right direction.

The point ahead where all this work had begun was located almost on the banks of the westernmost arm of the huge salt Lake Hopkins. It was the closest spot that the road came to the lake which in the original survey had caused so much trouble, with its maze of mud areas connected by long arms over which our road could not possibly be made. I had first stumbled into it from the south while endeavouring to reach the Sir Frederick Range by the shortest route possible. Although the scant information on the only map compilation available showed the rough position of the range, only a vague impression was given that there was some sort of obstruction in the vicinity. A blue smudge labelled Lake Hopkins was on the line of my plotted bearing to the range from the last vantage point seventy kilometres away, but no details of its shape had yet been obtained. At this stage aerial photography for mapping had not extended out far enough to cover this region, but the latest maps now show the lake as a sort of dismembered jigsaw puzzle, each piece representing a patch of bottomless salty mud. To make it worse, each of these is linked to its neighbour with narrow stretches of similarly soft mud, and once you are entangled in the mesh, it becomes a long and laborious feat of trial and error to be rid of it.

I had been struggling up from the south at right angles to my plotted course, in the direction of the next potential trig site at Sir Frederick, and had travelled over about twenty kilometres of bad sand-ridges by the time I finally cleared them. Now, I thought happily, an obviously low-lying area lay ahead. Then the first mud patch covered on top by a wafer-thin layer of salt appeared, extending to the right and left of my direction. Which way should I turn to clear it? I had had so many brushes with these potential traps that the Rover almost shied away from them by itself.

A little to the right the edge of the mud patch seemed to curve upward and away, whereas on the left most of the bank eased back from where I had come. This discovery caused me to make my first of many errors following on from the numerous trials. Keeping the rim at a healthy distance, I traced out the shape past the extremity and continued on to the north. Seemingly clear of the obstruction, I decided that Lake Hopkins wasn't so frightening after all — until I ran into the next expanse of mud, just one hundred metres distant. Another decision had to be made, and again a slight veer to the right seemed to be the best way. Leaving the mud on the western side, I was soon able to resume my so-called beeline for the range. It took only fifty metres this time before I ran into the next salty swamp through some low shrubs which temporarily screened it from view, but a slight left curve cleared it easily.

My direction was becoming disturbed, but so far we could still make the road over these wheeltracks, I thought, as a long strip of salt not three metres across cut off my path. It was plainly joined to this latest patch, so I followed it further around to the left. In fact I followed it so far that soon I was heading exactly in the opposite direction. Then I discovered that it joined on to patch number two. To attempt to cross over it would have been akin to signing a death warrant, because if a complete camel wagon could sink out of sight (as did happen on a similar saltpan), my Rover would surely do the same. Long ago, while still in the process of learning about these traps, I had been in such bogs for many days on end, digging away almost hopelessly at the bluish mess in my efforts to extricate my vehicle. By now I had learned a very hard lesson.

It wasn't very long before I actually cut my own wheeltracks and knew that the last hour of slow toiling over the rough hummocks had been for absolutely nothing. After a day of fruitless scrub-bashing, still trying to skirt this most perplexing collection of impenetrable quagmires on their eastern sides, it was becoming clear that this wasn't the way to try to bypass the lake at all. If only I had known, the morass almost joined on to a string of such impossible saltpans which continued on to the east for 300 kilometres, ending in the huge Lake Amadeus to the north of Ayers Rock. Lake Hopkins formed the western extremity of this long rift valley, but as I battled away none of this knowledge was to be gleaned from any existing maps.

Finally it began to dawn on me that perhaps I might try passing this latest obstacle on its western side, for the simple reason that any other method of attack couldn't be worse than the one I had been attempting all day. Of course, as far as I knew, the maze in front could also have continued on indefinitely to the west. It was at times like this that I wouldn't have altogether refused the use of a helicopter. After an overnight camp alone in among the confusion of bogs, I pushed the little Rover back in a south-westerly direction, completely clearing the forerunners of the main saltpan. This manoeuvre was to cut my own wheeltracks from the south and veer them off in a westerly direction.

After fifteen kilometres of grinding over half-metre-high hillocks of sand heaped up around outcrops of samphire bush, I began to despair that I was ever going to be able to carry on with my northerly course to Sir Frederick. The bank of a long narrow arm forced me back to the south-west and further away from my destination than I was when first striking the trouble. The range was still forty kilometres away from my plotting at the astrofix I had read during the night and computed by the headlights of the Rover.

Not wishing ever again to turn to the east, I decided that there was nothing left to do but keep going for as long as it would take before the saltpans cut out, even if it meant driving fifty kilometres or more. An access by road to the far north had to be found, and with this thought I settled in for an all-out battle with Lake Hopkins. I never found out who "Hopkins" was, but whoever named this mess after him must have borne him a powerful grudge.

Now that my mind was made up I was no longer worried, and I met the challenge head on with a sort of happy determination to see which of us would win, knowing that it certainly wasn't going to be the lake.

Not long after that was settled, I was almost disappointed to notice the salty bank flattening out and curling up to the north. Maintaining a healthy distance, with the dozens of blind alleys to the east still fresh in my mind, I traced out the shape until miraculously it headed off sharply to the east. Could this be the extremity of the lake, or would another collection of saltpans take over just out of sight through the scrubs? If they resumed as they had for the previous two days then I would keep turning to the west until they did stop.

After hours of traversing due north, not another sign of the dreaded salt lake reappeared. It was only then that I was confident I had at last cleared it and the way was now open. On my return journey I would carry out the usual procedure which had governed the location of thousands of kilometres of new road already. That was to retrace my wheeltracks to the point where it had become clear the current obstacle was safely coped with and, abandoning the reconnaissance tracks at that spot, to continue on back to the present head of the road. On the way I would determine the best location for the road with an eye to the sand-ridges and rocky outcrops, always easing the direction in such a way as to converge on my outgoing wheeltracks. These would have been made with the same thoughts in mind, and once intercepted, a few branches over the ones leading into difficulties would be enough to block them off. In that way a set of tracks would then lead from the road to bypass the distant obstacles, whatever they might be. The Rover's winding route would be soon ironed out by the bulldozer and then further mapping programmes would be carried out together with my list of astrofixes to pinpoint the final location. These last were continually forwarded from my camp to the National Mapping Council. Sometimes, when good paper on which to record the information was scarce, and to give the boys in the offices something to talk about, I would as a joke send the figures clearly printed on the reverse side of labels from tins of bully-beef.

Immediately north of the lake, acting as a final barrier between the end of that particular reconnaissance and the Sir Frederick Range, another belt of sand-ridges loomed up. It was certainly not an impenetrable barrier but it was to cause countless kilometres of slow bone-shaking travel to discover a way through it suitable for a road. I could beat my way through with slackened tyres over the ridges and roughly keep my direction, regardless of rocks and potential swamps, but for this project my wheeltracks must substantially lie on the final graded route. In that way no time need be wasted between my periodical expeditions ahead when the actual bulldozing was taking place. I had to be able constantly to guide the machinery a kilometre at a time or even metre by metre, confidently knowing that we were headed in the right direction and over the best possible route. If it were to lead into blind alleys it would cause a great deal of work and fuel consumption, as well as useless damaging of trees and vegetation. The position of the bulldozer must be at all times the exact site of the final access.

With this approach we didn't waste more than half a kilometre of distance in all the 7000 kilometres of road my little party constructed. Once, just as the dozer was approaching a sandhill to start a crossing, I thought Doug was waiting for my signal when in effect he was continuing on. In the short time that elapsed I had found a valley which bypassed that particular ridge altogether, but when I had pushed back through the thick bush to tell him, I found that he had overshot the turning point. We reluctantly reversed the big machine several hundred metres and made the right curve, not before bulldozing a dirt barrier to block off the cut which was to be superseded. The chance of completely avoiding a crossing had been too good to miss, but needless to say Doug and I received months of reminders from Quinny who insisted that after that effort, the bulldozer should be put in a museum and the operator and surveyor in the zoo.

During the detailed reconnaissance involved in handling that sandy buffer between Lake Hopkins and the Sir Frederick Range I had received a great deal of unexpected help in the form of several days of heavy rain. Normally rain brings everything to a stop because the bottom seems to fall out of the country and all attempts at work result in a hopeless series of bogs. The rain certainly stopped the roadmaking for several days, but when the sand was wet it became possible to drive over dunes which would usually be quite impossible even to approach. I could and did use this literally and practically heaven-sent opportunity to find a way through this section, and during the several days that the camp waited up out of the mud on a stony rise, my Rover was never idle.

When the time came for the bulldozer to come down from its safe perch on the hill and tentatively try its weight on the lower valley between it and the next sandhill, the path right through to the clearing around Sir Frederick was clear in my mind. During this stage of work the terrain was still wet but not boggy, and once again the resulting road proved to be longer lasting, being more firmly compacted than the construction of previous weeks.

One day, as I was flashing the mirror to signal Doug over a sandy saddle, I received a reply from his throttle lever. We had a system of communicating with each other in this way, and several bursts of his engine called me back when something had gone amiss. On this occasion it meant that the steel cable on the lift winch of the dozer blade had broken, so almost without stopping I continued on past it to fetch Rex and the spare he always carried. Driving over the freshly cut bulldozer line I soon reached the graded section with the grader sitting in the middle of the road motionless. On reaching it I noticed that the engine was stopped, and there was no Scotty to be seen anywhere. Skirting around the immobile machine I carried on to the camp and there, on the new smooth surface, were Scotty's hobnailed boot tracks, heading in the same direction. Another kilometre or so brought me within sight of the figure plodding along the road, and when he heard the sound of my Rover approaching he sat on the windrow and waited for me. As I stopped for him to climb in, I thought that it was a little early in the history of this road for me to come upon hitch-hikers.

It turned out that Scotty had run out of diesel — a very rare occurrence. The day before, he and Rex had not topped up the tank, thinking that after the spell on the hill it would easily last till midday; but with the heavier going in the wet ground the big motor had been working against its governor since starting, and the fuel had gone through at a great rate. This was another case of something good coming of a former mishap: if Doug's lift winch had not broken, Scotty would have had a long walk.

Just before breaking through this maze to the Sir Frederick Range clearing, and thirteen kilometres short of the desert oak with the astronomical values on its blaze, I had marked another desert oak with an item of interest. Complete with aluminium plate and results of a sun observation for latitude, I informed all future travellers that they were now at the one hundred mile point on the Sandy Blight Junction Road since leaving its source on the Gunbarrel Highway. For those interested, the south latitude of this goal was punched on as being 24°09′47″. Again, since then and in these metric days, I would have had to put the relatively uninteresting "mileage" as being 161 kilometres. Although considerable advantages are to be had in the long run from the great metric changeover, I realised later that I had been saved a great deal of work on these aluminium tree signs. If I had been inclined to mark the even hundred kilometre points in this way, I would have needed to carry a lot more aluminium.

Now, as we gradually left Sir Frederick's range behind, we could only sight it from the crests of some of the higher sandridges. The road curled and twisted its way through a belt which had required as much of a battle as the one to the north of the clearing, but the surface was of a different texture because of all that rain. It was now late November, and as our train continued on its way it seemed to us impossible that rain could ever have fallen on this sizzlingly hot countryside. Soon, as the sun began to dip below the sandhills, it was time to be looking for another camp.

The spot I selected happened to be the one on which, as a form of diversion, we had decided to test our rifles when we had camped on that cleared square before. With rows of jamtins arrayed on the far windrow of the new road we all blazed away for hours to determine whose aim was the best. Paul joined in with my revolver which he fired only once, sending a tin spinning off into the spinifex, before promptly retiring at the height of his amazing success.

To explain to anyone later happening to find the resulting heap of brass shells: this was not the site of Ned Kelly's last stand, but merely one of the camps of the Gunbarrel Road Construction Party.

The rugged gorge leading to the Bungabiddy Rock Holes in the Walter James Range—the miracle of fresh, clear water in the middle of all that arid country

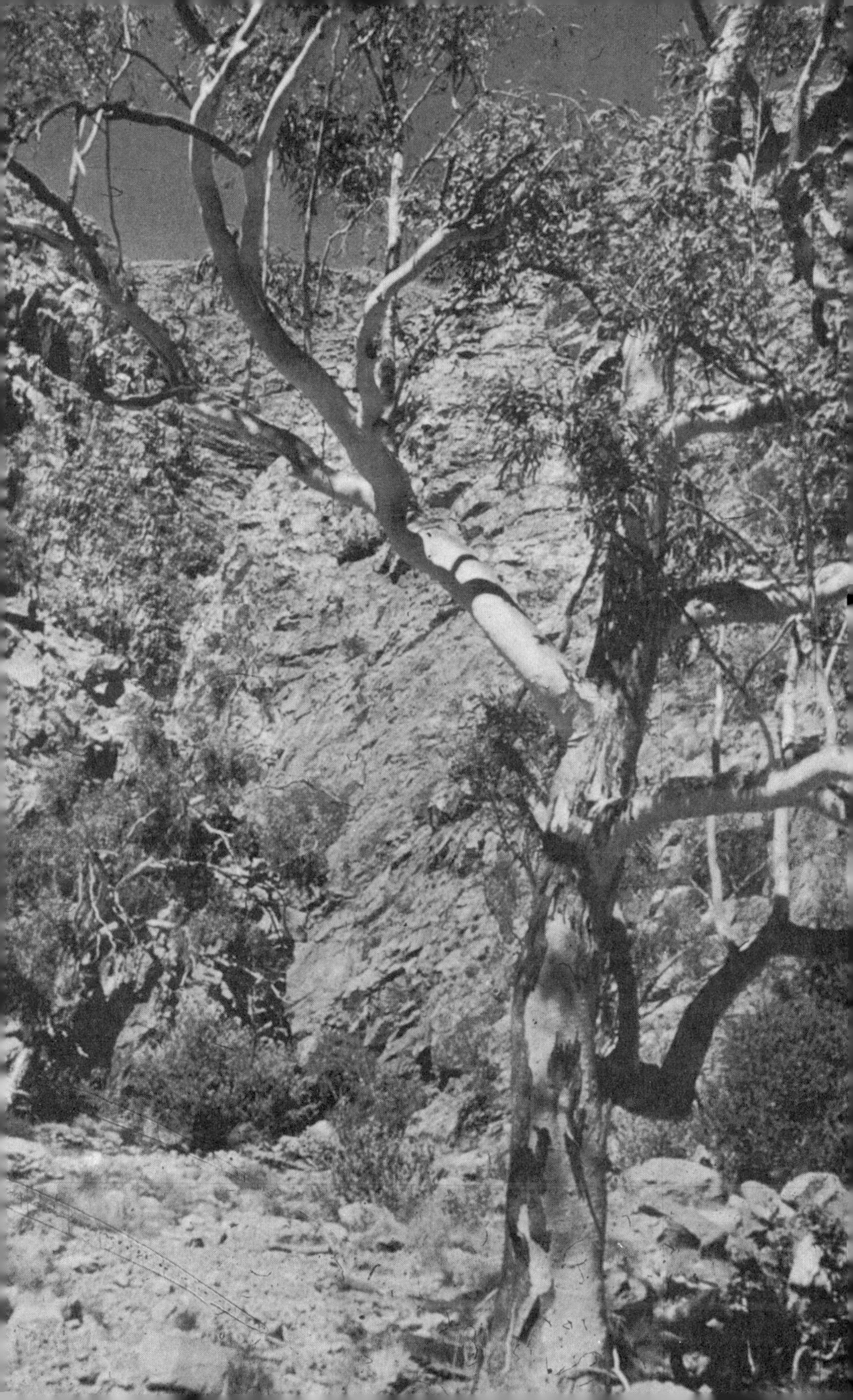

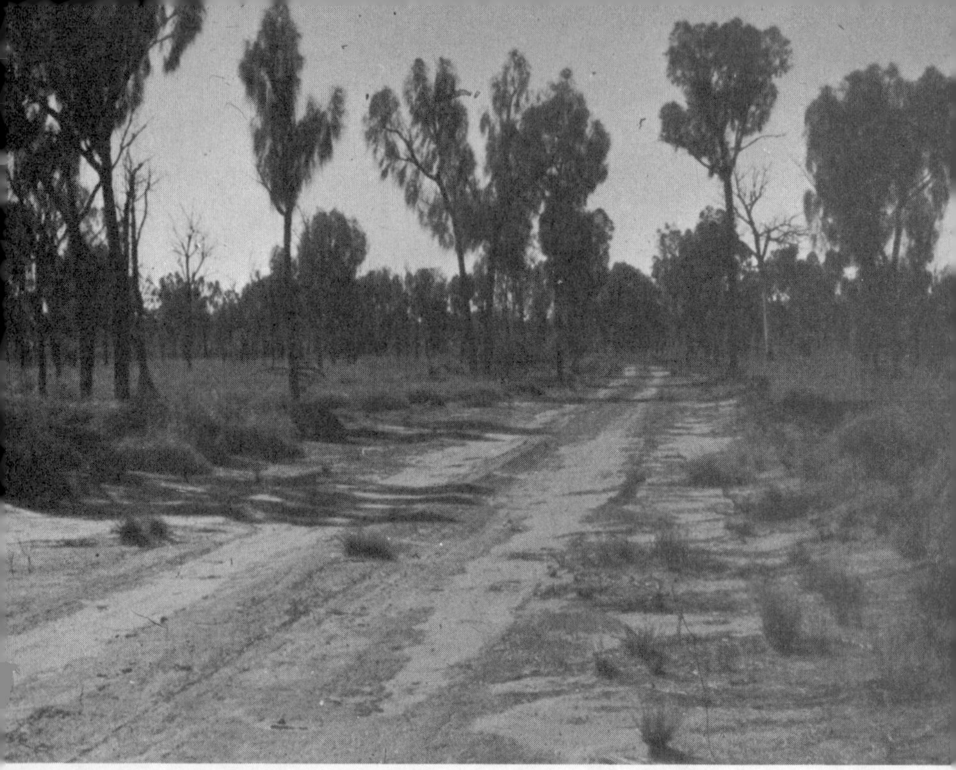

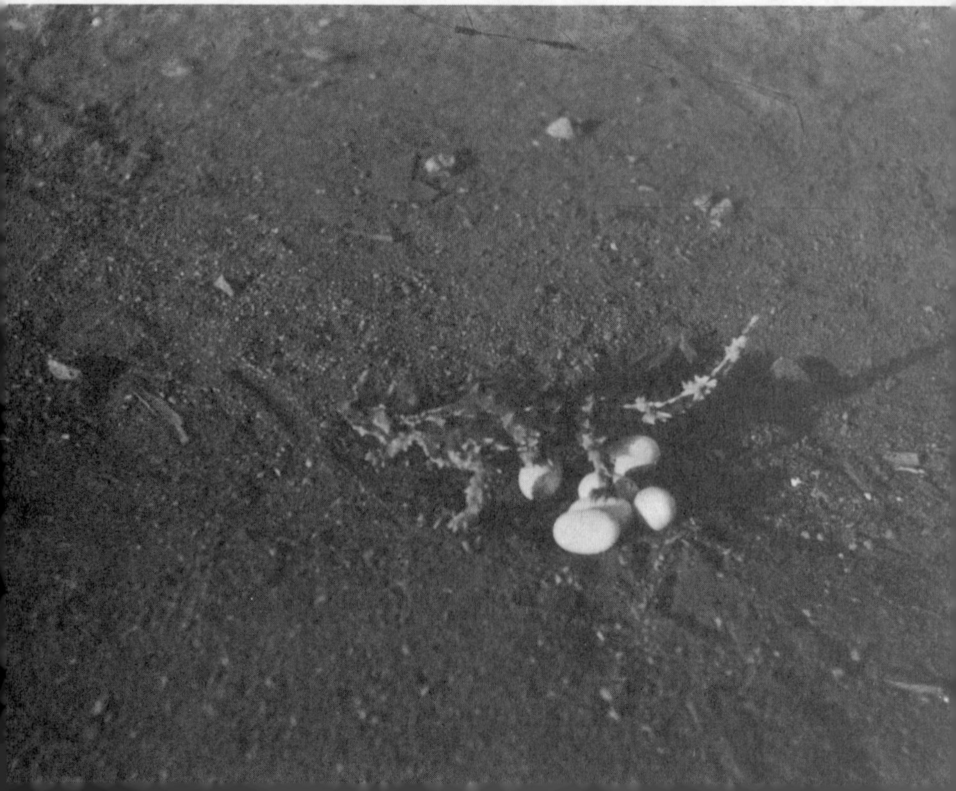

Chapter Fifteen

"THIS 'ARID' LAND"

As usual, the arrival of the next morning brought with it a renewed eagerness to be once more on our snail-like way. Today we should be over the last of the sandhill crossings to be encountered on this long haul and we should reach the westernmost bank of that terrible Lake Hopkins. If all went well, our next camp would put us within sight of the impressive spectacle of the red quartzite bluffs of the Walter James Range, which held for us a host of exciting and vivid memories. Thanks to the previously-mentioned absence of sandhill crossings, there would soon be a slight improvement in our daily travelling rate.

I had headed off towards Sir Frederick from our camp at Walter James on the direct bearing I'd calculated from an astrofix, when I became almost immediately entangled in sandhills. Being so used to the way in which these ridges continued on in either direction, sometimes for hundreds of kilometres, I struggled over the first one only to find a second and higher one waiting. After this a third, then countless others had to be attacked, making me doubt that any sort of a road could follow in my wake. Of course, at that time I was not to know that the position of Lake Hopkins would settle altogether any thought of a direct route, and in a way I was glad later that it had been there.

Top: We always tried to save the stands of desert oaks when we were roadmaking.
Bottom: A very rare sighting of a female Mountain Devil with a clutch of new-laid eggs

After my return reconnaissance from the westernmost rim I was able to run due south through a passage in the bush, absolutely clear of sandhills for twenty kilometres. All those high crossings on my way north had been avoided, as the ridges, instead of stringing out indefinitely, all came to an abrupt stop. As I was later to discover, they continued on to the west of this brief and unexplained clear passage for many hundreds of kilometres, but owing to the presence of the lake, I had been guided as if by some unseen hand towards my only northerly escape route. All that was required was to trace the shape of the first ridge in its adjoining spinifex valley until its terminus, and to turn towards Sir Frederick from there. All those countless hills stopped similarly in a line. I could have saved much fuel and wear and tear on the Rover, I thought, as I almost casually drove past them on the flat ground. It always came down to the one basic fact that it was easy once you knew how, but who could possibly know what this amazing country had in store without first expending a lot of sweat and toil to find out.

After about fifteen kilometres of travelling through and over the sandhills, which presented no problems now that the road was there, we emerged on to the actual banks of our old enemy Lake Hopkins, and while I waited for the train to catch up I looked with a mixture of fascination and satisfaction across the expanse of salty mud. There was no longer any need ever again to hold any fear of it with the road in place, easing up from the rim past a stand of desert oaks among which we had camped six months before.

When the train arrived we stopped for a track-roller greasing session, and it wasn't long before we were again on our way, heading south through the easy passage between the east and west extremities of the sand-ridges. Soon after leaving the lake behind, the time arrived for a dinner campfire, and I chose a clearing which had been the scene of an extremely rare occurrence when we had camped there before during construction. In all those years of continual camping in the open in Central Australia I had only sighted three other such phenomena, and on each occasion an indescribable feeling of eeriness had enveloped me.

It happened about two hours after sundown, as we sat by the fire endeavouring to absorb some heat into our frozen bodies before hurrying into our swags for the night. Without preamble or warning, a blazing flash in the southern sky lit up the entire countryside, momentarily converting the inky black environs of our camp into daylight. Trees, bushes, and equipment could be seen in detail as if on a sunny morning. The whole event lasted several seconds which seemed at the time an eternity.

I happened to be sitting at the northern side of the fire with an uninterrupted view of the sky around the Southern Cross, where this brilliant streak began and ended, and I was lucky enough to observe the whole episode. Scotty and Doug on the east and west of the fire saw the beginning only from the corner of their eyes but managed to see the end fully, and Rex used up half the time by spinning around to see what could possibly be causing the flash. Even old Paul and Quinny yelled out from the depths of their swags, demanding of no-one in particular, "What on earth was that?" When the excitement had temporarily died down and I was able to speak again, I informed the two prostrate figures that whatever happened was not exactly "on earth" at all.

The most baffling feature about this streak of light (as against others occasionally seen by those fortunate people out of doors at the time), and the one defying any attempt at explanation on our part, was the path it had taken. Usually such flashes of light in the sky take on a straight line in whichever direction the possible stray meteorite may be travelling when it becomes burnt up in the earth's atmosphere. This one, however, began near the skyline, then, blazing its way with ever-increasing luminescence in an almost vertical direction, it reached a culmination, turned over in a tight arc, and began a downward plunge. A third of the way down along the length of the upward trace, it abruptly trailed off in brilliance, leaving behind the shape of its trajectory lined in white. Because of our human "eye lag," the whole spectacle remained in place for a second afterwards. The colour of the trail was a lemon yellow at its brightest, and showers of sparks gave the impression of a length of bent hairy rope at the peak.

This happening of course settled the topic of conversation for the rest of that night as we sat in stunned awe around our relatively dull fire, each of us endeavouring to describe the event in his own words, and proffering his own version of the explanation. A straight line could easily have been written off as a burning missile from outer space hurtling through our stratosphere, and that explanation would have quite set our minds at rest, but a line with a tight hook in it and on such a gigantic scale defied our attempts at elucidation completely. It appeared for all the world like a child's Guy Fawkes night sky-rocket which, after attaining the height of whatever charge it held, began its return to earth to land in someone's fowl-house, scaring the life out of the inhabitants. What we had seen was a sky-rocket magnified to fantastic proportions.

I had seen another of these amazing sights ten years earlier, while I was reading an astronomical observation for latitude at the old atomic bomb area we had named Emu Field, 600 kilometres away to the south-east. The two events were strangely similar in that they were both exactly in the southern sky with their peaks in close proximity to the celestial south pole. Both had originated from the lower left and arced over to the right. I wondered how many other such phenomena had taken place when I had not been in the right place to view them, and if there was any significance attached to the fact that they revolved about the production of the earth's axis to infinity.

On that previous occasion I had been standing at my theodolite with its telescope pointed skywards, aimed at the star Beta Carinae just as it was approaching transit or culmination in the south, when the beginnings of this other streak between the standards of the instrument took my attention. All thought of the projected readings to the star disappeared as I gazed in wonder at the magnificent display before me. Nobody was anywhere near to share the thrill of it with me, as at that time the nearest person, black or white, was over 300 kilometres away. I remembered continuing on with my astrofix in a sort of daze as my mind relived the awe-inspiring event, and now an almost identical thing had taken place.

Part of the babble of discussion around our fire on this later occasion had concerned my description of the previous flash, and I was pleased and quite relieved that all the members of our little camp in the wilderness could vouch for this second experience. People were already beginning to consider that I'd been too long in the bush and I would have been reluctant to add the news of this sighting to the other, which was at the time treated with total disbelief. With my little group of hard-headed sceptics behind me I would have no hesitation in relating all the details of that memorable night of excitement, when and if we ever again returned to civilisation.

As I made up the small fire, the bulldozer at the head of its train protestingly squealed to a halt, and by the time the dust had settled the rest of the convoy were on hand for the lunch stop. As we all sat out in the blazing sun in this area devoid of shady trees and swiped at our accompanying mist of flies, we soon brought up the subject of what we had all seen on this very spot not really so long ago in terms of interplanetary missile activities.

Anxious to escape the flies and move on from this locality of intense bottled-up heat, Doug and I headed off as soon as Rex had the "locomotive" refuelled and serviced, leaving the rest to do the same when they were ready.

Less than ten kilometres further on brought us to the area where we had all huddled in our vehicles for three days sheltering from a bout of almost continuous rain, once more with the camp and our heavy plant perched on a stony rise up out of the mud. There had been no hope of moving anything off this platform, and even after the deluge had stopped several days elapsed before we could resume work.

During this period we tentatively tried to move off, firstly with the grader, because no matter how badly it became bogged, the bulldozer back-up on the stones would be able to pull it out easily — or pull it in half trying. We carried long lengths of heavy steel cable for this type of operation, and at our first attempt after the clouds had cleared away we had to make the maximum use of them all. Scotty drove off down the slope along the last cut made by the dozer before the rain hit and promptly sank to the chain-drive box between the rear tandem wheels. The momentum had put the front wheels down to over half their diameter in the mire ahead, but in anticipation of this Doug soon had those heavy steel tracks in reverse, straining on the cables which lifted the half-dozen tonnes of machinery bodily out and back up on to the hard ground. We wouldn't have entertained the idea had it not been for the presence of such power to rectify any mishap. I, being in a much lighter vehicle, was able to make use of the next few days to make a more detailed survey ahead.

Somehow Paul managed to continue cooking meals during this period, and we helped by dragging in chopped dry mulga trees to stack under the trucks out of the wet. A canvas awning attached to the top of the high canopy of the ration truck and anchored to half-metre-long iron pegs sledge-hammered into the rocky ground formed our dining-room. During the heavier downpours it also formed a protection over a smaller cooking fire. Paul's tea-chest library nailed to the wooden end of our ration chest on the back of his truck came in for its greatest share of customers at that time. Nobody and nothing was seriously inconvenienced by the spell: only the progress of the Sandy Blight Road construction was somewhat impeded. It had been rather bleak, but old army overcoats helped with that aspect. Nevertheless, we did miss the nightly roaring, cheery fires as these helped remove the bitter chill from our blanket rolls, into which we darted straight from the fire.

Now, in very different conditions, I waited on this burning hot rise for Doug's train. It first appeared as a shimmering dark mass, separated from the ground by the mirage which was ever present at this time of the year. All traces of the bog holes and evidence of the rain that had previously confined us to this ledge had long since vanished, but the road to the north remained in better shape for a number of kilometres as a result. Although it had received the same amount of rain, the section behind to the south, in the direction we were now travelling, had not the advantages to be gained by compaction as the weight of heavy vehicles was needed to compress the surface into a hard crust.

One day during this rare occurrence of such a long period of rain, the weather broke for several hours around midday, so I jumped out of the Rover to take a reading to the sun at transit for our latitude. I remembered setting up as the small patches of blue sky peeped through the otherwise full coverage of grey cloud, and as the sun shone through I began the observation. I had already computed the time of culmination, and the coincidence that it was to be seen just prior to the instant I had worked out put this scheme into my mind. With only a few minutes to go, the clouds began partially obscuring the sun once more, but by using a light-tinted eyepiece on the telescope I could still make out the orange shape. Now and then a heavier mass would hide it altogether, but between these times I continued the angles until the sun had gained its maximum altitude and was once more on its way down. Before I had put the instrument away in its box and the tripod safely back in the Rover, the rain began to fall, and by the time I was huddled back in the cabin, it was coming down again in torrents. I spent the next quarter of an hour happily reducing the results as a useful diversion from using Paul's library. The result told me that we were at present shivering in the dreary wet bush at south latitude of $24°32'$.

We were camped at a point on the upward trend in the road after we had been coming in an east-to-west direction between two walls of sand on either side for twenty kilometres. Those same ridges had been the first of hundreds over which I had flogged the poor Rover on my initial survey, and it was along this stretch during construction that several related incidents had taken place.

The first came upon us as we were sitting quietly having our evening meal out in the open. Heavy clouds had caused darkness to fall prematurely, giving us not much time to put up our awning. The canvas sheet served only to keep off heavy dew in those months and was only stretched out when time allowed. On the evening in question the sky had started to become overcast around midday, developing into sombre blackness by the late afternoon. We had sited our camp that night in one of the many low-lying depressions through which the road along this section passed, each of them relatively heavily timbered with mulga trees and separated from the next by open spinifex. The only source of firewood came from these patches, and on this gloomy afternoon the camp area had taken on the appearance of dusk long before the more open sections. The canvas cover was still rolled up and as we had just been through a long dry spell with many such false alarms, we didn't bother to stretch it out.

The bulldozer lay at the end of the road a few kilometres away, past several depressions, and Doug and I had returned to camp at the end of the day in the Rover. By the light of the roaring fire we all sat around our folding table with those on the far side in the freezing night air. Paul had given us our food and just as we had gratefully finished, a large spot of water hit the big aluminium salt-shaker with a loud clang. Several others followed in quick succession and so we decided to put the awning out and get the things under cover, for it looked as if actual rain was falling at last. No sooner had we begun the operation than the air was suddenly converted to water. The deluge came within one second, as though an overhead tank had suddenly burst. No warning other than the few spots had been given and the bright firelight had made the sky a black void, giving us no indication as to the density of the clouds. The rain must have been on a level plane as it fell earthwards, and the result was manifest in the way everything was washed off the table while we were still thinking of moving it. The ground was instantly centimetres deep in water. We were all soaked to the skin before we quite realised what had hit us, and as we were caught so completely unawares, our open swags and equipment suffered a similar fate. It was the most sudden torrent of rain that any of us had ever seen in our lives.

The big fire, which usually survived small showers, was visibly shaken and although it valiantly tried to radiate cheerfulness, it soon became a black simmering mess. Not much comfort was to be experienced by any of us that night, but at the same time we were cheered by the fact that the whole episode was over almost as suddenly as it had begun. Water cascaded in rivulets to the lower ground, taking with it jam-tins, plates, mugs, and anything else which had been on our dining-table. The next morning brought bright sun in a clear blue sky, innocently smiling down on our sodden camp as

Scotty attempted to sprinkle salt on his breakfast and was rewarded only by a shower of salt water over his plate.

The second episode occurred on that same day after we had done our best to carry on with the road in the waterlogged valley. Earlier that year I had a small winch installed on the front of my Rover, hoping for the day I next became bogged so that I could try it out instead of shovelling mud for days as usual. By skirting along the fringe of the sandhill on the north and around the quagmire at the centre of each depression we had managed to reach the dozer, but as it was hopeless to shift camp it remained where it was for the next few days. We repeated the process to return to camp, and so by lunchtime we had a deep set of wheeltracks safely up on the higher sand.

That afternoon, as Scotty could not attempt the grading so soon after the flood, the small workshop Rover came forward to collect Doug at the end of the day while I carried out a reconnaissance in front. On my way back along the ungraded new road, and with the safe feeling that I was now in possession of a brand-new winch, I decided to attempt to follow the road over a depression instead of going around it as we had been doing. The previous night's deluge had been enough to convert the new and uncompacted lower ground into a bottomless bowl, and on reaching the middle, down went the Rover, luckily on the passenger's side. Had it been on the driver's side I could never have opened the door with the mudguards half submerged in mud.

At last I could test out my beautiful new winch. It was only a small unit, but I had been assured that it was very well geared and extremely powerful. Jumping out and feeling pleased to have this chance of extricating a vehicle so simply after all those years of labouring with shovel and axe, I waded around through ankle-deep mud to prepare for the task. After first taking out my huge coil of crisp new rope from the back I joined it to the winch bollard and strung it out for fifty metres in front. The road had carved its way through the dense mulga which grew over the lowest points of the terrain.

The ground on either side of the road was as soft as the centre of the depression, although no actual water could be seen. Still, I was able to plod through the mud, sinking into it to beyond my ankles with every step. Wading back to the Rover, I engaged the levers on the incredibly miniature winch, hoping it would prove strong enough to do the job.

After half an hour, small as this machine was, I discovered to my amazement that it was powerful enough to drag out by the roots every one of the mulga trees in sight, leaving me to take my axe and shovel and carry on as usual.

An hour later the workshop Rover came into view, well up out of the scene of my downfall on the sandhill, and I returned to camp in that, leaving my faithful vehicle together with its wondrous winch to put in the night half buried in mud. We would get it out next morning. Darkness had already fallen by then and I was reasonably sure that no further torrential rain would arrive to allow it to sink deeper. The stars were already shining down with all their wintry brilliance.

It seemed a fair sort of a statement that Rex made to the assembled party around the renewed fire that night when he voiced an opinion which was very probably in all our minds. He made his comment that it was typical of the dust-dry, dead heart of Central Australia, after all he had heard about it, that it only seldom ever seemed to stop raining.

Chapter Sixteen

AN EXCITING THOUGHT

Throughout the whole of this current operation, and even more so now that the end of it was in sight, my mind constantly returned to the one thought which, although often pushed into the background, was nevertheless always present. It concerned the continuation of our latest road destined to cross the continent and how best we could handle it after our plant had been repaired.

Eight hundred kilometres of unbroken horizons of the spinifex-clad sand-ridges comprising the Great Sandy Desert lay between the abrupt end of the road and its ultimate goal. We had planned to bring it out on to an existing station track a scant sixty kilometres short of the mid-point on the Eighty Mile Beach washed by the Indian Ocean. This track led to Callawa homestead near the Oakover River, a further ninety kilometres away to the south-west. Our decision was based on the pattern of the sand-ridges as depicted clearly on aerial photographs covering the area.

A narrow passage formed by an apparent easing of sandhill intensity was oriented roughly along a north-west/south-east axis. This extended for nearly 200 kilometres back into the heart of the desert, bounded on either side by an extraordinarily dense maze of dunes, and was obviously far too good a topographical feature to ignore. The fact that it didn't lead direct to Callawa was completely incidental, as by then a way would have been opened across Australia to a spot which would provide access to the whole of the Western Australian road system.

All that was left to make this a reality was for us to battle our way through the intervening 600 kilometres of desert, leaving a road behind us. If the general direction of this could be aimed at the easternmost end of the open corridor the greatest advantage could be made of it and the last 200 kilometres should be comparatively easy. According to aerial photographs which had been made over parts of the Great Sandy Desert, that next section of country separating us from our goal was certainly something to be reckoned with, being completely covered by parallel sand-ridges. One feature was in our favour, however, and that was the general trend of the pattern. The countless ribbons of narrow valleys barely separating the ridges all lay in a rough west-north-west direction and I had already concluded that, once we were in a groove, it would be virtually impossible to escape from it.

Over that distance from our present road terminus, we would have to make up a 250-kilometre difference in latitude to raise us to the level of the end of the long clearing. Following the slight upward bearing of the ridges would only raise us about a third of the way, and it would be up to my old Rover and thousands of kilometres of reconnaissance to push the road northerly whenever the opportunity presented itself. Knowing where to go was one thing, but achieving or maintaining a direction in that area was something else again.

Although the long spaghetti-like strings of sand as they appeared on the photos seemed to continue on indefinitely they did in fact sometimes come to an end. Another would be right there to take over, but these temporary cessations would be enough at least to enable us to slip into the adjoining slot, for the spinifex-covered ground in the gaps was on the same level as the floors of the valleys. This I knew to be true from many years of handling such country.

Sometimes one of these ridges, far from petering out, merged gradually on to its neighbour, and the resulting combination was a disaster. If we happened to find ourselves in one of these blind valleys and discovered that the sand barriers on either side of us converged to block the way completely after what could have been 100 kilometres of roadmaking, then that entire section would have been wasted. It was my job, on my reconnaissance trips, to ensure that this situation never occurred.

In normal mulga scrub country, and even in rocky range outcrop areas, a route still had to be carefully planned. After clearing each obstruction—whether it was a salt lake, a gypsum hill, a confused sandhill belt, or rocks—a further survey would be needed to handle the next section. Gypsum hills (which are composed of the soft white powder from which plaster of paris is made) sometimes blocked our way with trailing flat areas of apparently bottomless chalk dust, but these were at least confined to definite areas. Once they had been avoided, a fresh series of unrelated problems would be waiting. This applied equally to saltpans and rock conglomerations but not to those endless sand-ridges which lay waiting for us in the Great Sandy Desert.

We had already negotiated other sections of country with comparable topography, including the 700-kilometre stretch of the Great Victoria Desert. The difference there was that the thirty-metre-high sand-ridges were densely covered with a tangled mass of mulga scrub. The grooves ran for hundreds of kilometres, but on that road at least, the east-west pattern almost helped us, considerably limiting the need for astrofixes. Even if I wanted to change course the mountainous proportions of the ridges, once started, totally dictated our direction, and it was up to us to begin in the right valley in the first place. Knowing that a year's work would be going into that road I had a long hard look at it first, but from then on it was mainly a slow grind with no end in sight for many months at a time.

At least our future project lay over open spinifex country, despite the persistence of the sand-ridges. Already the germ of an idea was forming in my mind to put this feature to our advantage. The utter remoteness of the area made it a gigantic operation to transport a bulldozer out to our starting point even before any progress could be attempted, but a grader, which can drive itself like a huge truck, could quite easily be brought out to the scene of activity. The biggest question was, perhaps, the absence of dense scrub: could the grader handle the job by itself? It was an extremely powerful machine and had already been used, without any help from its close friend the bulldozer, to make up the formation of parts of our other roads where they crossed open spinifex plains. Coping in this way with thousands of kilometres of new ground would mean four times the volume of careful survey, but it was well worth considering. Nowhere else could the idea even be contemplated; but as I sat waiting for our slow-moving train to appear, the thought of trying it out in the Great Sandy Desert was decidedly exciting.

In our overall scheme, the mere fact of extending the road west from Sandy Blight Junction to the Marble Bar area had another objective in view. Its completion to provide the first north-west passage across Australia was also of paramount importance because the path for which we would be striving lay virtually along the centre line of the proposed firing of long-range missiles from our Woomera Rocket Range. When it was finished, any instrumentation involved could be linked to the launching pads and the planned impact area would be accurately fixed as regards bearing and distance. As well as this, the tremendous programme of national mapping in this area would be helped considerably. For several decades there had been conferences aimed at bringing about an overall network of geodetic and topographical surveys resulting in accurate, up-to-date maps of various scales covering the entire continent of Australia. If the road project had been designed for this alone, it would have been amply justifiable in the rewards to future development; but it needed a project as unique as Woomera to initiate it.

The other ramifications to our particular share of these end results included at least two further major road connections. As these were also destined to straddle the range centre line and to thread their way through positively the most remote outback country to be found anywhere in Australia, they were equally important and needed to be brought into our future planning.

The first of those remaining involved a 350-kilometre connection between the Callawa road (when it was made) and our original now well-established Gunbarrel Highway. This almost due north-south road was to be situated in the absolute nucleus of the Gibson Desert which in itself had no definite boundaries other than covering "all the dreary wastes west of the Rawlinson Ranges." As this isolation extended west for 750 kilometres to the nearest habitation, with nothing but dry sandhills and spinifex all the way, our projected road link would be in a geographical location literally screaming to be recorded. There again some aerial photos had been taken, and from an inspection of those available, a conspicuous lessening in the intensity of the sandhills could be detected. It appeared extremely likely that after a thorough on-the-spot appraisal of the area, resulting in a planned programme of reconnaissance expeditions, a road could be made to by-pass the core of the maze of dunes. That task also might possibly be handled by the lone grader. The appearance of the area was of open, spinifex-covered undulations with heavy belts of sandhills both to the east and west of a north-south band stretching for most of the way and separating the two major roads.

These natural features seemed to be heaven-sent, which in a way they probably were, to aid us in our work. Such a road system as the one we hoped to finish up with could not have been attempted if it hadn't been for them. There was admittedly a distance of about seventy kilometres laced with sand ridges towards the northern end of this particular planned road, but I was sure that with many solo trips of exploration I could turn the otherwise straight line direction into a shape that would eventually form a junction with the as yet non-existent Callawa road. What did give me cause for concern, for it was something I had already come across when we were planning and making the Gunbarrel Highway on the southern end, was the belt of mulga scrub through which we would have to pass before arriving at the start of the open undulations. Normally, if we were going to use the bulldozer, this would not even have entered my mind; but how might it affect the feasibility of carrying out this new scheme with only the rubber-tyred grader? Every day, as our towing operation progressed, I felt more and more strongly that it would be unwise to bring out a bulldozer to this most remote of places if the rest of the road could be achieved without it. At this stage I had no real knowledge of the extent of the mulga belt, but perhaps after my first survey I would be in a better position to assess it. It wouldn't be impossible to carve out an access laboriously around the trees if they didn't persist for too great a distance. I dwelt on the fact that only three years before, right on the very spot where I planned to tap the Gunbàrrel Highway with a junction for this new road, we actually did have a massive bulldozer. If only I'd known I would have included that lead-off at the time, but it had been impossible to provide for such details so far in advance.

In any event those questions were yet to be answered by endless and exacting survey trips in the years to come. It almost seemed to me that this enforced slow towing operation might have been wreaked upon us on purpose in order that we could have a breathing space to contemplate our future moves while the obvious problems were still so fresh in our minds. I don't expect any other member of the party gave it even a fleeting thought as we ground our way to Giles through the

One of the water-filled rock-holes discovered in the Rawlinson Ranges by the explorer Ernest Giles. Mount Destruction can be seen in the distance

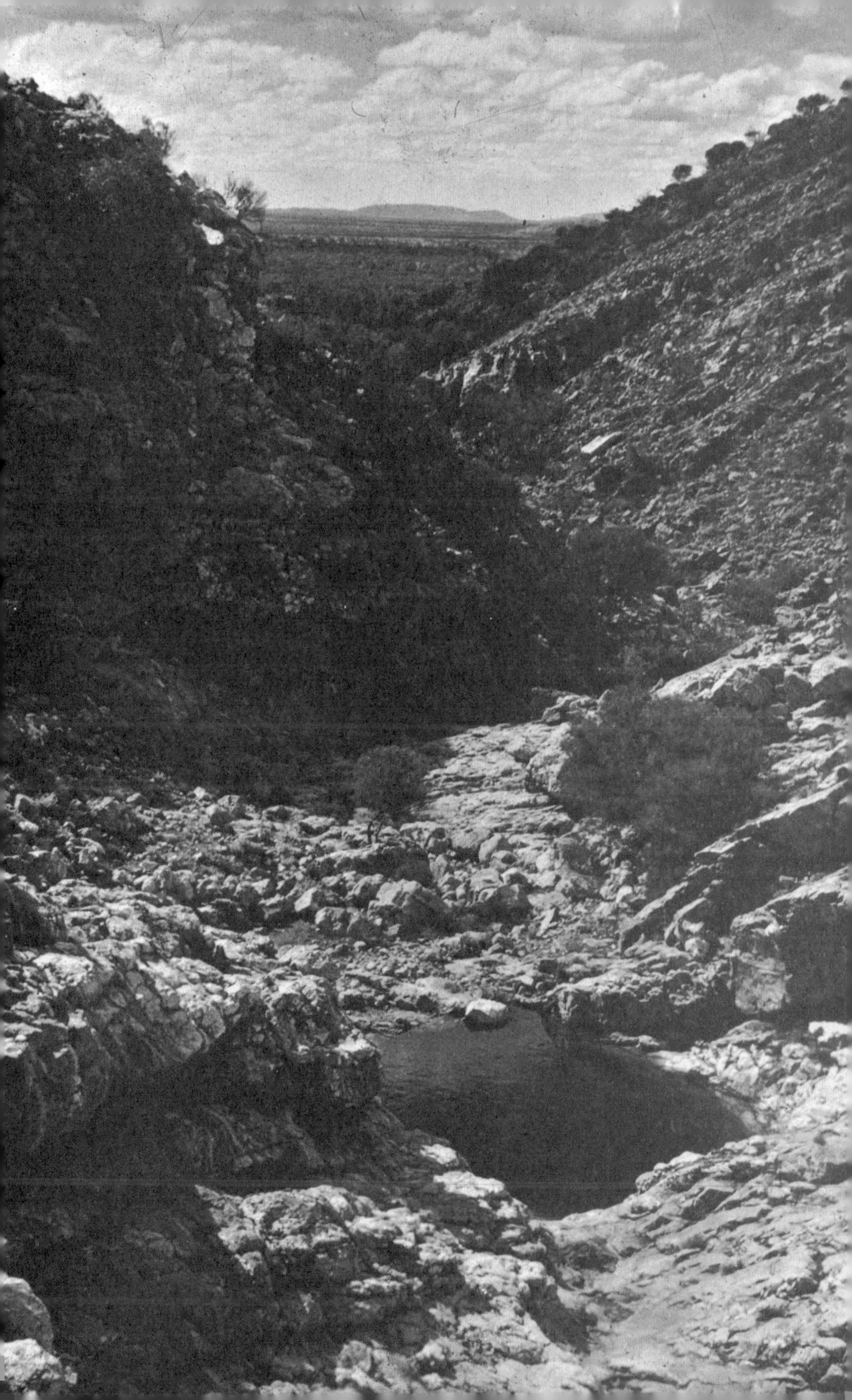

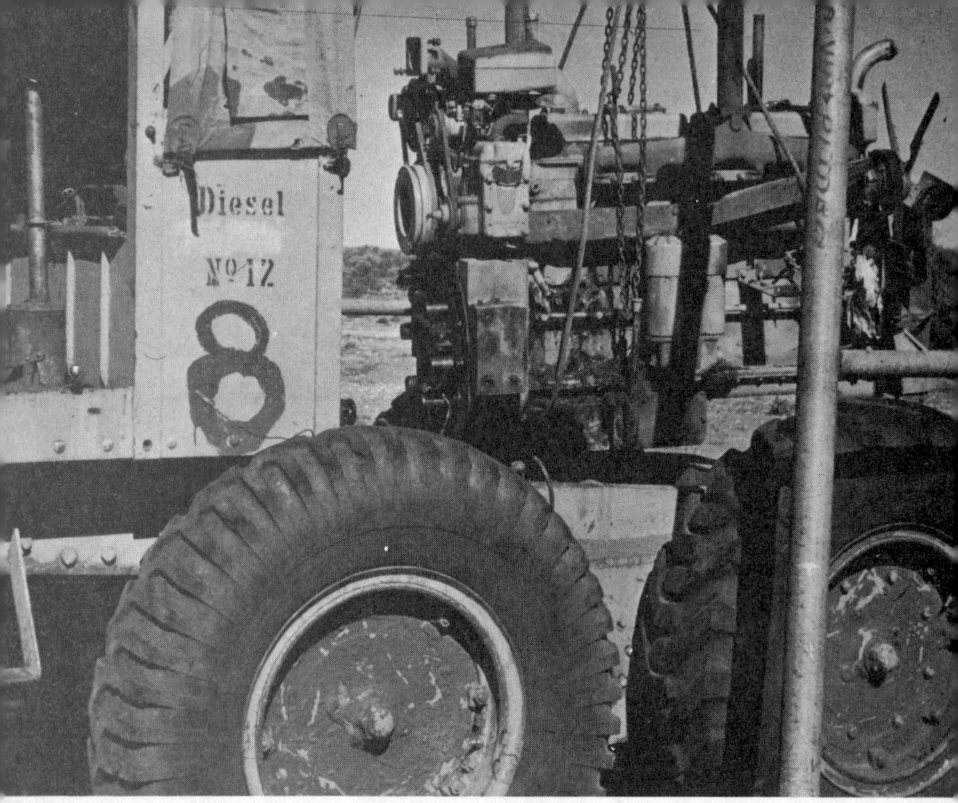
Diesel
№ 17
8

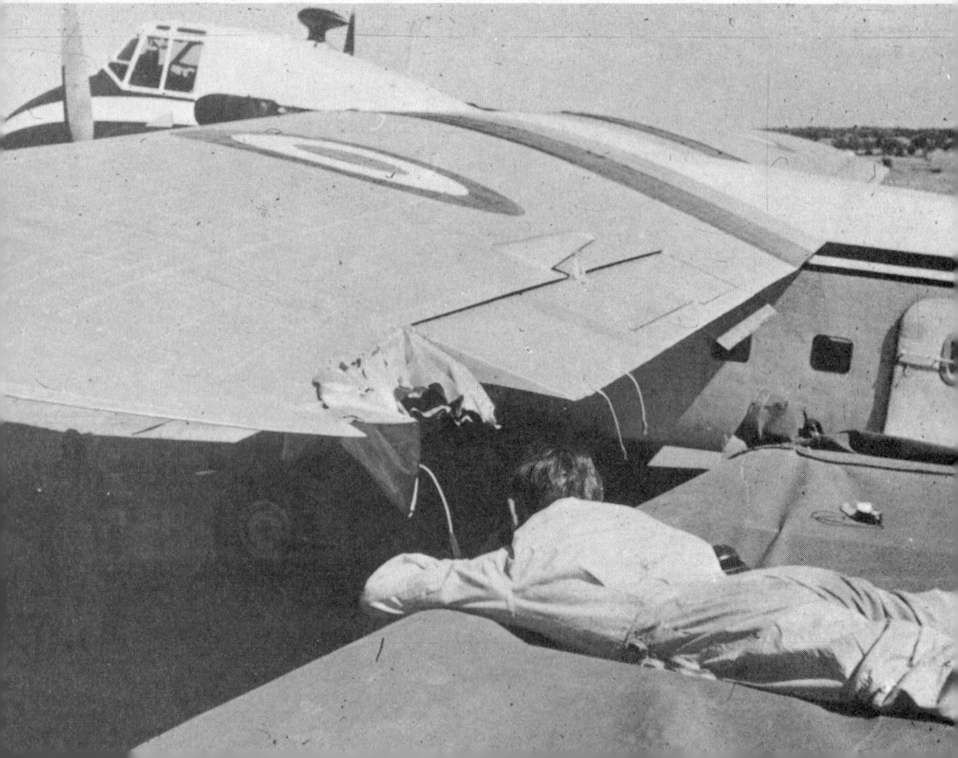

inferno of early summer on the desert, but at least I could brood about it to my heart's content. The exploding truck, tooth extraction, windmill mending, and camp preparing only served to interrupt my one main concern: the best way of tackling our new work next year.

Assuming the road to Marble Bar and the connection through the Gibson Desert finally could become an actual reality, there was yet another 500-kilometre road to add to our score. All the remaining area of the Gibson Desert roughly bounded by our three bush highways and to the west of the north-south link was still entirely inaccessible by any sort of a road. The quarter-million square kilometres of country would need to be penetrated by the follow-up survey parties, and I could see by merely glancing at the void plotted on a map of Australia that a road would simply have to be made through its centre. The shortest length to bisect it would need to be about 500 kilometres, and a road running from east to west would provide the greatest advantage.

According to the otherwise almost blank map, a large salt lake covering 2000 square kilometres would appear about halfway along the length of this as yet hypothetical road. Lake Disappointment, as it was labelled, was probably one of the most remote dry saltpans in Australia, and as it lay almost at the centre point of the remaining desert to be penetrated, I concluded that this leg of road would have to skirt it. At this stage of course my vague locations for the other sections were only pencil marks trailing over the bare paper.

The incredible stock route planned by Surveyor Alfred Wernham Canning and Trotman, his second in command, skirted this lake on its western side as it travelled in a general north-easterly direction from Wiluna on its way to Halls Creek 1500 kilometres distant. The course of the route was studded at intervals with wells at an average distance of twenty-five kilometres apart and numbering fifty-two. All had been fixed by Canning and Trotman using astronomical observations, and were plotted on the only maps covering that country. The route, completed in 1910, had been instigated as a means of droving cattle from the Hall's Creek area in the north of the State down to the railhead at Wiluna. A problem involving

TOP: A typical open-air workshop scene. Repairs on our heavy machinery were usually done on site, out in the bush. BOTTOM: "It was a good aeroplane when we landed, but look at it now"

cattle tick meant that the beef could not be disposed of westerly to the coast and as a result mobs consisting of 500 head of beasts were driven south-westerly, taking about five months in the process. Hard-bitten drovers and leathery bushmen, camping in the toughest country to be found anywhere, pushed the cattle along month after month from well to well where they were watered before continuing.

The Canning Stock Route cut squarely across my two projected westerly roads, and as we pushed our plant down the Sandy Blight Road, I couldn't help likening our operation to that of the drovers. Converting all our equipment into its own weight in cattle and substituting stockwhips for throttle levers, the distance travelled was at least the same.

Tracing the course of the pencil line on the location of the northernmost future road to Marble Bar, well Number 35 on the stock route was located at a rough plotted distance of just over 200 kilometres from the head of our current road. I knew that I would make the new road to bypass it within a few metres. It would have to be found first as easily as finding the proverbial needle in a haystack, then the flow of the road would be made to it. There would still be over 500 kilometres of road left to reach Callawa but it would greatly increase the interest at one point along the way, as well as perhaps providing a possible supply of water to travellers. The quality and quantity of water in these wells was unknown because I had gathered the stock route hadn't been used for several years. (In fact it was never used again after about 1959.)

My planned road west from the north-south link through the Gibson Desert also cut across the Canning, again 200 kilometres on its way, at Lake Disappointment. As the stock route curled around to the west and north of the great saltpan, I noticed that there was a short east-west section of it with a well at either end. What a great point of interest could be added to our road if I could make it pass by *both* of them, and how grateful some later prostrated wanderer might be to be guided to water in that desolate region. Even if his need for water was non-existent, the thrill of actually seeing anything different from sandhills and spinifex in such a featureless waste would make the slight diversion well worth our trouble.

Again, these wells would be difficult to find as since the cessation of the cattle drives the stock route for the most part would be completely invisible, but I had the comforting thought that their plotted position was from actual on-the-spot star observations. I would bush-bash to the area, fix my own position from the stars, and calculate the bearing and distance to them from there. By a series of such astrofixes combined with sun observations, I knew I could at least work out their plotted positions before searching until I pinpointed them.

The two wells north of Lake Disappointment were labelled Number 23 and Number 24, with well 24 also being called the Karara Soaks. From any of my starting-points an ocean of sandhills separated me from these areas, and the task of working through them was still far in the future, but already I could hardly wait to begin it in the New Year.

From the sorry appearance of our gear at the present time, it would seem to an onlooker that if and when we crawled it into Giles, then it would all die there. But we had been in straits almost as dire as this many times before and had resurrected the equipment to carry on with our work to date, and I knew that we would definitely press on again with our exploration and access roads into even more remote areas than ever before.

From what few aerial photos were available to me I had been gleaning that as well as the long sand-ridges lying in a certain overall direction, there was an added feature to the pattern. As the ridges went further south, forks began to be formed in the system. Long trailing spurs joined to their neighbours while others separated from those spurs to add yet another prong. The consistency of this trend caused me to give careful consideration as to my future direction of attack in the initial surveys I would be doing before we could begin any actual plant work. Because all the points of these prongs seemed to be on the eastern ends of the ridges and sand spurs, a reconnaissance towards the west would be constantly thwarted. More often than not, after settling into a valley to laboriously trace it to its conclusion, I would find myself bottled up as the adjoining sandy barriers converged as one in front. I had often been confronted with this situation when making the Gunbarrel Highway, and drawing on my previous experience I concluded that the only way to lunge into this Lake Disappointment region of the Gibson Desert was from the western edge. In that way my expedition tracks might flow on past the prongs uninterrupted by blind valleys and I could discover where best to lead the road to its western conclusion. We could then start on the point on the north-south link where I emerged from still another long solo expedition.

My mind went on and on with plans for the methods I could employ to make all this a reality and at the same time see the resulting roads laid down where they would be of the most benefit to everyone both at present and in the future. At that time, as for ever afterwards, I felt closer to Ernest Giles and Alfred Canning than to any other earlier explorers and bushmen in the history of Central Australia.

It was while camping between the Robert and Walter James ranges on that sweltering November night by the side of our still too-hot-to-touch machinery that I had the first glimmering of an exciting plan.

What if we didn't carry on with our road *from* its present temporary terminus at all!

Say we were to proceed along the Gunbarrel Highway instead, to a point where the junction for the north-south link would best be created, and construct that first. Of course at that stage it wouldn't actually be linking anything but a spot on the future Callawa road. After forming another junction, we could then make the connection south-easterly *back* to the head of the road we had just vacated on our towing operation, joining it head-on. If all this could be accomplished with the grader alone, it wouldn't be a difficult task to backtrack to the second junction and carry on right across the Great Sandy Desert via well Number 35 to Callawa. That would put me on the western end of the third projected road to carry out the survey past Lake Disappointment and via wells Number 23 and 24 to hit the centre of the north-south link. The plant could return by our brand-new road from Callawa to meet me if I got through.

It was all a decidedly exciting prospect. As sleep finally claimed me, I knew I had a job of unprecedented and titanic proportions awaiting me.

Chapter Seventeen

BUNGABIDDY AND HOME

Within half a day of pulling out from our overnight camp between the sand-ridges east of the Robert Range, the train emerged from the valley. As we rounded the last of the dunes, the Walter James Range was at last in full view. The Robert Range, located ten kilometres west of that historic upward turn in our road, was named by Ernest Giles in 1874 after his brother and was for us the last main topographical feature before Walter James.

On my solitary reconnaissance to discover a break through to the north, I had camped relatively close to the Robert Range in a deep basin completely ringed by high steep sandhills. After the battle with Lake Hopkins I came upon some rocky bluffs, a little too far to the west, as the memories of those salty swamps had been fresh and I had been endeavouring to give them a greater berth than was necessary. These vertical cliffs were named the Wallace Hills and the huge belt of impassable sand-ridges bordered them on their western side. Of course at the time I hadn't known this. I was unwittingly about to find it out the hard way.

Beating through the scrub and sand until the sun had long since set, I decided before camping to tackle the one large ridge in front of me at the time in the gloom of dusk. After a dozen attempts I felt elated at finally winning and slid down the other side to the flat valley in a cascade of sand, knowing full well I could never retrace my tracks back in that direction. I couldn't see further than a few metres, but once on the level surface, I soon found among the spinifex a clearing on which to settle down for the night. I needed a clearing as I had two flat spare tyres to mend, but I decided against the need for an astrofix on that spot.

Too tired after the day's bush-bashing and repairs after dark to think of eating, I lay down in my swag over a hundred kilometres from the rest of the boys waiting at the current head of the road at Walter James. Before long — or so it seemed — daylight had begun to approach once more in the form of a faint glow over the dunes to the east. Up and anxious to be on my way I rolled my blankets in the semi-darkness of the hollow as the sky became lighter by the minute, and looked about me. Ahead lay a high and very steep sandhill. This curved around to the north to connect to the huge one I had slid down in the darkness of the previous evening, barely a hundred metres away. Tracing the outline to the south, I was amazed to discover that it followed an almost identical course, arcing right back again to the one behind. I was in a deep basin completely surrounded by a circular wall of sand, with every avenue of escape apparently blocked off. Searching in vain for an outlet in any direction, I realised that without knowing what I was doing in the dark I had landed myself in a drastic predicament by allowing the gallant little Rover to be carried by the moving sand to the bottom of the dune. I was so far from the nearest living thing that any possibility of walking was out of the question, and I knew straight away that this was going to mean a lot of hard work if I was to ever see my camp again. Even if I did manage to emerge from this bowl, there were still 100 kilometres of as yet unknown country to penetrate before I could reach the rest of the party, for so far on this expedition I had only succeeded in discovering an impossible route through Lake Hopkins. The final location for my road was as yet an unknown quantity. With this in mind and with the daylight by then almost fully upon me, I studied the details of my amphitheatre.

At one point to the south-east I noticed that the barrier was slightly lower than it was for the remainder of the entire circle, and I concluded that my exit route would have to be over this section or not at all. As I approached it, my hopes faded, for the potential crossing seemed to increase in height with my closer proximity. Nevertheless I selected the appropriate gears on the Rover and tried. That first attempt barely lifted the vehicle over its own height from the floor of the arena and I knew then that this was going to get worse before it could get better. Backing the Rover in a straight line well clear of the pathetic little progress of that first effort, I climbed out and deflated all my tyres to a point beyond which they would fall off their rims. They had already been slackened to a stage which made it possible for me to get into this dish in the first place but that clearly wasn't enough for this do-or-die operation. I considered unloading and dragging each item over the barrier on foot to lighten the vehicle, but decided that often a little weight adds to the traction if any. This plan would be kept for a last-ditch effort. The thought of a further method waiting for me always goaded me on and raised my spirits sufficiently to tackle the one in hand.

Using my little shovel, I started work on making a road in the same way as the early convicts in Australia built them. The only things missing were the leg-irons and the harsh boss with his flogging whip. Smoothing the hummocks of spinifex and replacing the dissected tussocks evenly on the route, I worked for hours to attain a smooth, if steep, grade to the top where initial momentum might carry me over. How I longed for the bulldozer resting quietly back in my camp with its huge gleaming blade in front of those great Caterpillar tracks.

At long last, and with the sun by then beating down on me as I worked, I had managed to construct a twin set of tracks from the front of the Rover to the summit of the ridge. Then, looking hopelessly around me at the circle of even higher dunes, I clambered into the driver's seat. This attempt brought me a third of the way and left a good straight set of tracks. Replacing the disturbed spinifex and smoothing out the small hillock which had built up in front of the wheels at their furthest point, I repeated the process to gain another metre.

With renewed hope and after fifteen more attacks I knew that the next one should clear the crest. I had already climbed to the top to ascertain if I had an outlet on the other side, or if another such basin awaited me there. If that had been the case I might have been able to remain on the razorback crest and follow it around, but after that early inspection I found that all was well.

Carefully replacing the moved herbage back on to the tracks, which by this time were quite deep, I returned to the vehicle only to find a front tyre absolutely flat on the ground. The torture had proved too much for it in its slackened state. Thankful for the previous night's mending, I replaced it with a newly-patched "spare" and sorrowfully allowed all the air, which I'd laboured to pump in by hand only hours before, to escape. If this last charge was successful I would gladly reinflate all the tyres as I couldn't afford to waste petrol by using the engine plug pump.

I don't think anyone has expected more from a motor than I did when the vehicle was within a metre of that virtually unattainable crest, but the length of run-up and the gradual consolidation of the deep wheeltracks won over, and with the engine screaming at full pitch, I found myself on top. I didn't lose momentum until the whole weight was at last on the downhill slope, when I stopped and trudged back for my shovel. I hadn't dared pack it away or take it upon myself to assume success until it was positively achieved.

Now, with those Wallace Hills only a memory and with the Robert Range receding to make way for the Walter James Range, I had an increasing feeling that the worst of our long haul was over. The spectacle of the red quartzite cliffs held a much more friendly atmosphere for me. It had been here that I had finally returned to camp after a successful expedition on which I had discovered a route through almost impossible country for this part of the Sandy Blight Junction Road.

I had first set eyes on this range on a similar journey to survey ahead a course for the road which at that time reached only to the easternmost termination of the Schwerin Mural Crescent. Between there and the Petermann Range further to the east existed a complete break in the range system which extended from 140 kilometres to the west through the Rawlinson Range. The Petermanns took up almost as soon as the Mural Crescent left off, and the Rebecca Creek, also named by Ernest Giles, threaded its dry course between them.

The presence of this pass had led me to decide on a starting point for my new road north into the Northern Territory. I came to this decision only after much laborious work testing a route through the only other western outlet at the Giles meteorological station. We had established this weather station on the southern side of the Rawlinsons only six kilometres from this break (named by Giles as the Pass of the Abencerrages), and I endeavoured at the time to continue my road through it. After several long hard expeditions during which I constantly ran into barriers of rocky outcrops and dense sand-ridges I was forced to the conclusion that such a route was not easily possible, so I decided to make use of the only other outlet to the north at the Rebecca. As it turned out I have often been glad I did, for the quiet little station set in its most pleasant surroundings would have been continually beset by plagues of future travellers, each expecting help in one form or another.

With a view to people one day coming up from the south on my original road to the station, I selected a point which made most use of its northerly trend before curving westward to form what is now known as the Gunbarrel Highway, and created a junction. A pleasant stand of desert oaks thirty kilometres short of the station made an ideal camp for our initial base on the existing road, and this has now become a junction well known to countless travellers.

With the new road and my camp at its head on the banks of the Rebecca, I had need of a detailed knowledge of the country to the north to guide the road, so, as on so many occasions before, I plunged off alone into the scrub ahead. The north-westerly trend of the Rebecca governed the first dozen kilometres of the route because I didn't want the main access to cross it, and immediately after leaving the watercourse I became entangled with the rough rocky outcrops of the Anne Range. Fresh from failure at discovering a route through the other pass, I wondered if such a road was to be feasible at all;

but happily and after days of exploration small outlets among mazes of sand belts appeared at the right places to admit the width of a road.

After clearing this rocky obstacle I pushed through the thick bush once more on a northern course, seemingly for the time being unmolested by topography in this completely unknown section of Central Australia.

Soon I could see another mountain range towering above the comparatively diminutive one I'd just negotiated, and in ten kilometres of heavy scrub-bashing I felt I was level with it, although still five kilometres distant. I thought that the extreme troubles I had encountered close to such huge outcrops for almost every kilometre of progress to date must fade as I moved further away from them, but at the same time I often stood on the Rover's roof with binoculars to study this newest example. After all, the road was initially meant to provide access for a trig survey for Woomera, to help with the mapping of areas hitherto uncharted by nature of their extreme remoteness, and trig and tellurometer stations could only be established on hills giving line of sight between each other. The closer the road was built to high outcrops, the easier it was for the following survey parties.

Late one afternoon, as I scanned the eastern face of this Walter James Range through the glass, with the low sun's rays highlighting prominent features, an exciting deep black crevasse held my undivided attention. I knew instantly that I must go to it, so I jumped from the roof of the Rover and turned the vehicle towards it. In the cabin it was immediately obscured by scrub, but after an hour of beating the gorge could be seen at ground level, only half a kilometre distant. My feeling of excitement mounted as I contemplated the imminent visitation of such an isolated and remote feature. The depth of the crevasse could not even then be ascertained, and it looked more intriguing by the minute.

Finally, with my further progress blocked by rocks fallen from the towering cliffs after eons of weathering, I abandoned the Rover and continued on foot with the noise of my hobnailed boots echoing eerily in the otherwise silent canyon. Within half a kilometre, by moving in the only possible direction allowed by the vertical walls, the waterway elevated me to a level ledge across the entire width. Although the heat was intense at that time of the year, I quickly climbed the last few metres to the top of the stone barrier.

I could hardly believe my eyes at the sight spread out in front of me. A triangular-shaped smooth rock basin a dozen metres across had been formed naturally in the cleft between high cliff walls, and it was brim full of crystal-clear rainwater. The clear bottom receded into the depths at the centre. Hurrying down to the water's edge, I scooped up handfuls of the purest liquid I'd seen for months and drank gratefully. Only minutes before I'd been forcing my way through kilometres of dense hot dry bush, covered as I always seemed to be in a mixture of sweat, dust, and spinifex gum, and here in this hidden gorge lay an oasis too perfect to conjure up even in dreams. After so long away from plentiful water, and feeling as I did, the urge to plunge into this cold pool overcame all thoughts of preserving it, and soon I was diving into it from a rock ledge above, trying to reach the bottom at its centre. After several attempts I found I could not dive deep enough to feel a floor and concluded that this must be an almost permanent supply. It would be filled after even a small shower once or twice a year, catching the quick run-off from the hundreds of square metres of smooth rock all around. The sun's rays would seldom reach it, and with this depth at their disposal, Aborigines for countless generations would have survived the longest droughts.

Climbing further up the gully, I found that yet another and even bigger and better pool was located at the base of a most spectacular unscaleable fold in the rock formation. By now daylight was fading, and I was suddenly brought back to earth, realising I had a long way to go if I was to rejoin my camp that night. I couldn't easily reach my old shorts and hobs resting on the far bank of the first pool without a last plunge and swim over to them, and soon afterwards I was on my way back down to the Rover. As I clambered over the rocks I decided that wherever else it went, our road would be located as close as possible to this spot. With that in mind I drove off only as far as was necessary to clear the foothills before turning south to my outgoing wheeltracks.

Back in my hot, sweltery camp I sprang out of the simmering vehicle and informed the small group that I had just been for a swim in an almost bottomless stone tank, after diving from a high rocky platform. But nothing I could add would make them believe me, and I might as well have tried to convince the dusty hot mulga scrub around us that I hadn't completely taken leave of my senses. All I was rewarded with was tolerant smiles of agreement and a friendly word from Paul that of course I'd been swimming in the desert. "What else?" Quinny was for once without comment and retired to his swag on the water-tanks in his truck. Scotty gave his opinion that intense heat often does this sort of thing to people, while Doug just ignored me politely. Eric followed Quinny's example and escaped into the caravan. I smiled smugly to myself and said inaudibly, "You'll all soon see."

So it was that a week later our new road skirted the lower slopes of the Walter James Range, and with a view to obtaining water, even though it would have to be hand-carried to help our meagre supply, we bulldozed a short access from the main road to the entrance of the gorge. Only then did the party begin to have a suspicion that something might be there after all. Clearing a small area with the grader against the actual mountain the camp settled down for the night; and next morning with a flourish I led everyone, including Paul and Quinny, up to my discovery.

My credibility rating with the boys as a result of this episode lasted for months after this memorable contact with the Bungabiddy Rock Holes, to give them their official Aboriginal title. My standing seemed to wane at the Mount Leisler area but the unusual collection of huge red rounded boulders there served to revitalise it. These things contributed to the story of our lives during our eight years together on our monumental project.

As our bulldozer train lumbered on past that little offshoot into Bungabiddy, I, and I'm sure most of the others, relived the time eight months before when the Sandy Blight Junction road had progressed no further than that spot. Now here we were, almost the first to make use of our own road, limping back from the far-off deserts with what was left of the equipment that had made it. In one way it was as if it had dug its own grave. But we all knew that once we were back at Giles, it would all be repaired ready for our fresh onslaught into the bush in the new year.

That night, after clearing Bungabiddy and the Walter James Range, we camped again on our old area by Rebecca Creek, right between the two main range systems. Scotty, Quinny, and I even found a water soak in its bed. At that time of the year and in this area, the sight of water other than from hidden rock-holes was remarkable and meant that some little shower had fallen—though we certainly hadn't seen it.

The Schwerin Mural Crescent which now reared up alongside us had been also named by Giles after the Princess of Schwerin, who was married to the Emperor of Russia. Along its length was Vladimir Pass, named after their son. However he thought of such people when naming these features completely eluded me as I couldn't think of any connection whatsoever. More to the point was the name of Rawlinson given to the main range to the west, after Sir Henry who was president of the Royal Geographical Society in London.

The next day, as our train groaned along, still protesting in vain at the long drawn-out effort, we slowly passed the beautiful ghost gum alongside which we had carved the road, with the first of our aluminium signplates for this section installed on the blaze. Having no distances as yet to add, and no need of astronomical values for latitude and longitude, it merely gave the names and vocations of our party, the date when the new road had reached the spot, and the usual note added to inform future users that the Gunbarrel Road Construction Party was to blame for it.

In the background, a few kilometres away and standing remote from the main crescent, the curious formation of Gill's Pinnacle reared up abruptly from the scrub-covered level ground. Giles had named this after his nephew—a rather closer relation than the Emperor of Russia's wife.

The next overnight camp on this incredible and unprecedented towing operation was, almost sadly to say, to be our last. For almost a month this slow progress had become a way of life for us, and although it had been a gruelling experience, we were somehow sorry to see it coming to a close. I selected for our final stopover the same site among the desert oaks at the junction where we had camped when we began this whole year's work, mainly as the timing would be right after the distance from the Rebecca had been covered. Another reason was that I planned to erect a signpost to advise newcomers of the ultimate destination of our brand-new road, together with other relevant information, and I would be on the spot to complete it.

In anticipation of this I chopped a long straight length from one of the magnificent stands of oaks which at the outset we had reluctantly been forced to clear seventeen kilometres on our way. Lead capped, the post should last for many years, I thought, as I attached the plate which I'd spent an hour happily punching out, and painted the whole thing the usual white.

During that night, as we lay spreadeagled on top of our swags in the oppressive heat, endeavouring to gain the greatest possible degree of relief from a gently moving air current, I had cause to think that this was not only to be my last project, but my last night on earth. Through the fog of a restless attempt at sleep, a crawling sensation across my throat brought me back to cold reality. Motionless, I felt this slow movement across my bare shoulder and over my neck. I was quite sure that a deadly spinifex snake was making its lethargic way over me, and that one flicker of movement on my part would send poison fangs sinking into my pulsating veins. I lay there for what seemed like hours, not daring to even swallow. The incessant creeping persisted until a weak increase in the breeze caused an immediate corresponding flutter across my chest, transferring the contact from my throat. At that instant I thought it was now or never. I leapt up in what must have been the most rapid movement I had made for years, and looked down on my moonlit swag. There, alongside my grimy pillow, lay the old ragged shirt I had shed prior to collapsing for the night, with a willowy shred of material surging slowly about at the will of the breeze.

Paul was the only one who ever heard about that, as the others would never have let me forget it.

Seven hours after moving off saw our desert cavalcade at the foot of the stony rise upon which we had sited the weather station a half dozen years before. Once on top Doug manoeuvred the train to an out-of-the-way position and kicked out the master clutch-lever for the last time that year. Then, climbing down with an air of great importance, he threw down the decompression lever and that faithful big diesel engine died immediately.

Eric made for the bore-water shower room while Paul unceremoniously headed for the kitchen to help the station cook cope with meals for the swelled numbers. Quinny was all set to turn around and return to the desert again, a feeling I shared with him, and Scotty joined Rex to inspect the still simmering train whose dust had barely settled. We had come 800 kilometres from the end of the new road, which right then terminated at an inconspicuous mound of spinifex deposited by the bulldozer before turning around.

As if to soften the traumatic blow of this most abrupt

cessation of the lifestyle which had motivated us for nearly a year, several incidents occurred in quick succession to occupy our minds in the following few days.

Firstly, one member of our group came to me with the separate pieces of his upper denture which had somehow not stood up to the pounding in his shirt pocket as he did his washing; so once again out came the little dust-proof box. The plate had to be repaired quickly, as he wasn't to be left out of the enticing new food available. Then, seeing that the box was open with its contents spread out, one of the meteorological observers thought he might as well have a temporary filling installed as well.

The second excitement came the following morning when an Aboriginal woman limped in with a mangled toe for treatment. It was an obvious case of gangrene, and this meant a long radio conversation with the doctor at Woomera. Following the detailed instructions relayed to me, I performed an amputation, surrounded by Aborigines clutching bunches of spears. I wrapped the toe in paper and returned it to her, as it was hers anyway, together with the opinion that if she put it in a glass of water overnight, the fairies might pay her a visit.

Then, as if to completely erase any lingering depression at being in comparative civilisation again, if only temporarily, a Bristol Freighter which was due with supplies landed on the near-by airstrip that we had located and constructed ourselves when starting Giles, only to be crippled after a high-canopied station truck drove into its wing-tip. The pilot made the somewhat heated but quite unnecessary statement that it was a good plane when he landed but now it was useless.

One thing was certain: the Sandy Blight Junction Road would never be forgotten by our little party, if only for the fact that we had completed what just had to be one of the longest and most arduous towing operations in the history of Australia, if not indeed for the rest of the world.